RESEARCH PROBLEMS IN BIOLOGY

INVESTIGATIONS FOR STUDENTS

SERIES 1

SECOND EDITION

BIOLOGICAL SCIENCES CURRICULUM STUDY

New York
OXFORD UNIVERSITY PRESS
1976

FOREWORD

The *Research Problems in Biology* are designed to provide the superior biology student with a means of developing the art of investigation. The series originally appeared in four volumes. The wide acceptance of these books indicated a need for continued availability of such opportunities for potential biologists to exercise their ingenuity.

Since the introduction of this series in 1963, both the research climate and the emphases have changed. Many of the problems initially included are no longer applicable, and new biological concerns have arisen over the past decade. Therefore, it seemed appropriate that the series be revised to reflect the biology of the 1970s. To this end, each of the authors who had contributed prospectuses originally was invited to evaluate his contribution in terms of its current applicability and its value to potential biologists. As a result of these invitations, certain prospectuses included in the first volumes have been eliminated, a few have remained essentially unchanged, most have been revised and updated, and a number of new ones have been added. In addition, the arrangement of the research prospectuses has been organized into six themes that reflect today's biological concerns. Thus each of the three volumes in the series presents the student with problems in behavior, physiology, ecology, microbiology, genetics, and growth, form and development. Each grouping is defined in a broad sense.

The research prospectuses herein have each been written by an expert as a guide to scientific investigation. They are deliberately open-ended and relatively unstructured so that the individual investigator can exercise a maximum of individual participation in his selected project. The large number of suggested investigations help to insure that each student will find a topic that appeals to him, and one that he can master. The problems not only concern themselves with a variety of biological subjects, but they offer varying degrees of challenge to accommodate a wide range of biological sophistication. From our experience with the first series, we know that these research outlines are used for every kind of student endeavor, from science fair projects to doctoral theses. Their use is limited only by the abilities of the individual investigator, the degree of sophistication he brings to the project, and the resources and time at his command.

FOREWORD

The purposes of the individual investigations, then, are to present the outlines of real research problems at the forefront of today's biological efforts—problems to which there are no known solutions and relatively little literature. The investigator is largely on his own in exploring these fields of biological research, and the investigations may very well test his resourcefulness. He may even solve one of these perplexing problems and experience the justifiable pride of accomplishment.

Addison E. Lee, Chairman of the Board
William V. Mayer, President
The Biological Sciences Curriculum Study
P. O. Box 930
Boulder, Colorado 80302, U.S.A.

INTRODUCTION

Over forty scientists have joined to bring you this book. These scientists share with you the conviction that the study of a science is not only a matter of covering a vast body of information and making it one's own, but also of uncovering new knowledge as well. It is hardly necessary to emphasize that to work in science one must know what scientists have been doing and what they are doing now. Yet the core of the scientist's way of life is *inquiry.*

To an extent, the past achievements of scientists are recorded in all kinds of books and periodicals. In books you will find the common stock of knowledge, the body of information, confirmed and reconfirmed over and over. One who would search for new knowledge needs to possess a reasonable amount of this "common body of information" if he would achieve ordinary literacy in the field. At the same time it is well to understand that as a result of inquiry this "common body of information" changes in character simply because scientists are constantly acquiring new knowledge.

These forty-plus investigations were developed to bring you the opportunity to gain experience in the art of scientific investigation. You probably will not find "answers" in textbooks to the problems these investigations pose, nor do we expect you will find more than a possible avenue to their solution in the references appended to each one. However, with careful thought and zealous work, the imaginativeness and inventiveness you bring to the investigation may yield new knowledge—new technique; you may be the first to know something no one before you has known.

How does one begin an investigation? Begin at the beginning; this seems obvious. Precisely what aspect of the problem that you selected appeals to you? Try to make a statement of what you want to investigate. Then search the references; the library is one of your basic tools of investigation. Read carefully and searchingly what others have done. In the light of what you have read, and in the many talks you will have with your instructor, you will come to an understanding of some of the prime things an investigator must deal with. Are you, for instance, personally equipped to deal with the problem? Do you know enough to state the problem so that you can design a clean-cut experiment or program of observation?

Fortunately, an experiment tests itself; it often goes wrong when poorly designed. This happens most often when the problem is not clearly stated or understood at the beginning. If you have read carefully, you may be able to predict the outcome of the experiment; your hypothesis will have been tempered by the facts at your disposal.

Science, Einstein has said, is an experience in search of meaning. Experience, however, can end in frustrating failure if one is not prepared. This is not to say that failure is not to be met; indeed, failing intelligently is a great skill of the investigator. Each failure is analyzed, and from the analysis success may be fashioned. We hope, therefore, you will have a reasonable number of failures. There may be times when things just won't go right. If anything can go wrong with an experiment, it usually will. How well you face your problem when it goes sour, when nothing seems to go right, will be a test of your ability as an experimenter, but it will also furnish you some of the joys of the experiment—the overcoming of challenging obstacles. After all, you are on a road where probably no one has been before. Soon, however, if you persevere, you will find yourself working along new lines, asking fresh questions and designing fresh experiments.

You will have help if you seek it properly. University scientists will help. Industry will help. You have but to seek help to find it given willingly. Sooner or later you will have acquired a great gift—the confidence to work with independence and courage.

ACKNOWLEDGMENTS

It is a pleasure to express thanks to the authors who have supplied new research problems, and to those who have revised old ones. A most cordial thanks goes to the National Council for Biology Teaching in Mexico for the use of several problems.

Research Problems would not have been possible in either of its editions without the planning and direction of three key scientists and educators: Bentley Glass, formerly Chairman of the Steering Committee, BSCS; Arnold Grobman, formerly Director, BSCS; and Paul F. Brandwein, formerly Chairman of the Gifted Student Committee, BSCS. Acknowledging their continuing influence upon science education is a recognition they richly deserve.

Dorothy S. Curtis
Consultant, BSCS
Editor of the Series

CONTENTS — SERIES 1

Tables of contents from Series 2 and 3 appear in Appendix A to this volume.

BEHAVIOR

ANIMALS

MICROORGANISMS

CONTENTS

CONTENTS

CONTENTS

GROWTH, FORM AND DEVELOPMENT

ANIMALS

PLANTS

1 / THE INFLUENCE OF LIGHT AND TEMPERATURE ON THE ACTIVITY OF FISHES

Background

M. A. Ali
Department of Biology
University of Montreal
Montreal, Quebec
Canada

Animal activity is a result of the interplay between the organism and the physicochemical environment. Environmental factors such as light, temperature, pressure, pH, and chemical composition of the environment govern the rate of activity of the animal. For example, if a bright light is turned on over an aquarium containing a group of fish that have been in darkness or in moderately dim light, the animals tend to swim more actively. Although they quiet down after some time in the bright light, their activity is greater than it was while they were in the dark.

Conversely, if the animals that have been subjected to bright illumination are placed in darkness, their activity diminishes considerably. Many investigations have been carried out to show the role of environmental factors in the lives of animals (*see* Fry and Clark).

Recently, Ali (4) has shown the effect of light on activities such as feeding, schooling, and migration in the Pacific salmon (*Oncorhynchus*). In these experiments, the techniques were simple and the equipment necessary was inexpensive and easily available.

Suggested Problem

The main purpose of this project would be to observe the influence of two major and interrelated factors such as light and/or temperature on a particular behavioral response such as feeding.

Suggested Approach

It is suggested that fishes such as goldfish, guppies, medakas, or sticklebacks, which are easily procured and reared, be used. It is advisable to keep not more than two species of fish at any time and also to keep the species separate. A culture of *Daphnia* to be used as food in the experiments will need to be maintained. These water fleas can be collected quite easily from small ponds and maintained thereafter in the laboratory, using simple methods (3).

1

Three to five fishes should be trained (conditioned) to feed on *Daphnia* at a particular time every day on the application of a stimulus such as tapping on the side of the aquarium with a glass rod or pencil. The fishes will probably learn to expect food at this particular time after about two weeks' training. During the training period it is interesting to maintain a record of the number of *Daphnia* the fish feed on in a given time (5 or 10 minutes). This may be done by actually counting the number of captures the fish makes during that period or by offering 50 to 100 *Daphnia* at one time. At the end of the 5- or 10-minute period, remove the fish and count the remaining number of *Daphnia* in the aquarium. If the water in the aquarium is carefully poured away through a narrow mesh net, this net can then be immersed in a flat dish and the remaining *Daphnia* easily counted.

When the fish are conditioned to feed in this way, the actual investigation can begin. For the purposes of this study, you will need, apart from the fishes, *Daphnia*, glass aquaria, a darkroom (needed only for an hour or two on the days during which the experiments are conducted), a few cardboard boxes with apertures of various diameters, a light bulb and socket, a photometer to measure the light intensity in footcandles (G.E. pocket exposure meter is inexpensive and good) and a running cold water supply (for some experiments involving temperature). For more particulars of the methods, see Ali (4).

It is suggested that observations be made on the feeding rates of dark-adapted fish at various time intervals after their exposure to light. This would give an indication of the rate of light-adaptation of the animal. Subsequently, the feeding rates of animals under various light intensities may also be studied. During all these experiments, temperatures must be kept constant as far as possible.

Using the same method, the rates of feeding under constant light but under *different* temperatures might also be studied.

If the fish are maintained with care, it is possible to use the same set of fish for the entire series of experiments. Except for the day prior to the experiment, the diet of the fish should include other items of food, in addition to *Daphnia*. The diet should also contain Vitamin A (*see* Kampa).

In those experiments in which light intensity is a variable, the amount of light can be controlled by placing the light source in boxes with different sized apertures. This is a better method than using a variable rheostat, since changes in the current flowing in the bulb change the spectral composition of the light as well as the amount.

It might also be interesting to study the retinal responses of these fishes under the various conditions already outlined if facilities are available to prepare slides. For more details on histological techniques you may want to refer to Brett and Ali (7), Ali and Hoar (6), and Ali (5).

It is advisable to carry out experiments at the same time of the day instead of doing some during the morning of some days and some during the afternoons or evenings of other days.

Temperatures can be controlled by varying the temperature of the room in which the aquarium (with standing water) is situated or by placing the aquarium in a running tap-water bath and controlling the rate of flow of water.

Throughout the study, an attempt should be made to observe the behavior of the animals under the various experimental conditions to which they are exposed. See Tinbergen (2) for some interesting information on this topic. Quite often, observations made in this manner yield very interesting explanations to some of the results of the experiment.

References · general

1. Prosser, C. L. and F. A. Brown, Jr. 1961. Comparative animal physiology. 2d ed. W. B. Saunders Co., Philadelphia.
2. Tinbergen, N. 1951. The study of instinct. Oxford University Press, New York.

· specific

3. Ali, M. A. 1959. A suggested method for *Daphnia* culture. Turtox News 37:203.
4. _____. 1959. The ocular structure, retinomotor and photobehavioural responses of juvenile Pacific salmon. Canadian J. Zool. 37:965-996.
5. _____. 1960. The effect of temperature on the juvenile sockeye salmon retina. Canadian J. Zool. 38:160-171.
6. _____ and W. S. Hoar. 1959. Retinal responses of pink salmon associated with its downstream migration. Nature 184:106-107.
7. Brett, J. R. and M. A. Ali. 1958. Some observations on the structure and photomechanical responses of the Pacific salmon retina. J. Fish. Res. Bd. Canada 15(5):815-829.

8. ____ and D. A. Higgs. 1970. Effect of temperature on the rate of gastric digestion in fingerling sockeye salmon, *Oncorhynchus nerka*. J. Fish. Res. Bd. Canada 27(10):1767:1779.
9. Clark, G. L. 1930. Change of phototropic and geotropic signs in *Daphnia* induced by changes in light intensity. J. Exptl. Biol. 7:109-131.
10. DeWaide, J. H. 1970. Species differences in hepatic drug oxidation in mammals and fishes in relation to thermal acclimation. Comp. Gen. Pharmacol. 1(3):375-384.
11. Fry, F. E. J. 1947. Effects of the environment on animal activity. Univ. Toronto Studies, Biol. Series No. 55, Publ. Ontario Fish. Res. Lab. 68:1-62.
12. Kampa, E. 1953. Effects of Vitamin A supplementation and deprivation on visual sensitivity, rhodopsin concentration, retinal histology of a marine teleost, *Gillichthys mirabilis*. Bull. Scripps Inst. Oceanog. Univ. Calif. 6(6):199-224.
13. Umminger, Bruce L. 1971. Osmoregulatory role of serum glucose in freshwater-adapted killifish (fundulus Heteruclitus) at temperatures near freezing. Comp. Biochem. Physiol. 38A:141-145.

2/ THE SPECIFIC PREDATOR-PREY RELATIONSHIP BETWEEN THE MOSQUITO FISH, *GAMBUSIA AFFINIS,* AND THE PICKEREL, *ESOX AMERICANUS* OR *NIGER*

Background

Harper Follansbee
Phillips Academy
Andover, Massachusetts 01810

Dr. Carl George, in experimental work for his doctoral thesis at Harvard (June 1960), has discovered an apparently specific response of the mosquito fish, *Gambusia affinis,* to the presence of its natural predator the pickerel, *Esox americanus* or *niger.* Upon introduction of the pickerel into an aquarium containing a school of ten or more *Gambusia,* the following behavioral pattern of *Gambusia* emerges within 30 minutes: the iris darkens very noticeably and a dark suborbital bar appears prominently; the fish swim at the surface, ungrouped, axial musculature rigid, fins erect and unflexing; certain behavior of the pickerel will cause them to break the surface of the water. Certain parts of the pattern may appear in response to other stimuli such as the introduction of any large fish. The appearance of the total pattern and its persistence, however, appear to be specific to *Esox.* The actual stimulus appears to be chemical in nature since *Gambusia* will respond when placed in aquarium water in which *Esox* has been swimming for one hour prior to the introduction of *Gambusia* and from which *Esox* has been removed prior to the introduction of *Gambusia.* In addition, it is possible to remove the stimulus by filtering the aquarium water through activated charcoal. The stimulating chemical has not yet been isolated, nor is the means by which *Esox* releases it known. Also unknown are the location and nature (olfactory, gustatory, other?) of the receptors in *Gambusia.*

Study of this phenomenon contributes to our understanding of a specific predator-prey relationship. From a practical standpoint, discovery of the chemical structure of the stimulus might be beneficial to man in mosquito control, since the stimulus effectively drives *Gambusia* to the surface, where mosquito larvae and pupae are located during the mosquito breeding season. At a more basic level, study of this phenomenon may help us to understand better the way in which complex behavioral relationships evolve between two different species of animals. The behavioral pattern exhibited by *Gambusia* appears to have "survival value," since it has been demonstrated that the percentage of successful strikes by *Esox* is considerably higher for "bottom" fish than it is for "surface" fish, despite a considerably higher density of population at the surface.

Definition of the Problem

In addition to the unsolved problems mentioned above, there are many questions concerning this relationship that are as yet unanswered:

1. Dr. George encountered difficulties in obtaining the response from some varieties of *Gambusia* from different areas of the country. What is the relationship between the ranges of the two fish (which do not overlap completely) and the presence or absence or intensity of the response?

2. Do *Esox* from different areas of the country all secrete the stimulus chemical? Do both male and female secrete? Do other species of the Esocidae, such as the muskellunge and pike, evoke the response in *Gambusia*?

3. Does the quality of the water (swamp water versus fertilized pond water, high mineral content versus low mineral content, etc.) enhance or detract from the nature of the chemical stimulus?

4. Is there a possibility that the stimulus is actually visual and that *Gambusia* then secretes an "alarm" substance?

5. What is the importance of factors such as group, size, sex and age to the response?

6. Is the response, or parts of it, "learned" or "innate"? This will involve a real understanding of what is meant by these two terms and also of the role of maturation. Since *Gambusia* are live-born, naive fish could be isolated from older fish. In fact, Caesarean sections could be performed to insure the naivety of the fish both visually and chemically.

7. What is the nature of the relationship between the two species kept together over a period of time in a very large, especially constructed aquarium in which the natural habitat is closely approximated?

8. A possible field study for one living in the area where *Gambusia* and *Esox* are found living together naturally would be the effect of the relationship on the densities of the populations of the two fish.

Special Problems

Unfortunately there are no specific references on this particular problem, since Dr. George's doctoral thesis is as yet unpublished. Basic techniques and experiments on the problem have been incorporated in material compiled for a laboratory block on animal behavior and published by the BSCS (1). It is recommended that the student begin with this material.

References

1. Animal behavior. 1961. A laboratory block prepared by the BSCS Committee on Innovation in Laboratory Instruction.
2. Lorenz, K. Z. 1958. The evolution of behavior. Sci. Am. 199:67.
3. Scott, J. P. 1958. Animal behavior. University of Chicago Press, Chicago.
4. Tinbergen, N. 1953. Social behavior in animals. John Wiley & Sons, New York.

3 / REPRODUCTIVE BEHAVIOR OF NORTH AMERICAN STICKLEBACKS

Background

Wilbur L. Hartman
Bureau of Sport Fisheries and Wildlife
Sandusky, Ohio 44870

Some of the most interesting and fundamental research on fish behavior has been done in Europe by N. Tinbergen and his students on the three-spined stickleback, *Gasterosteus aculeatus*. Many of their results have added considerable insight into the broad study of animal behavior. Early detailed observational research was followed later by experimental research on the causal structure underlying the reproductive behavior. Fish models and other props were used to determine environmental factors that may act as sign stimuli and serve to release behavior patterns at different times during the reproductive cycle.

It would be valuable to study, in addition, the reproductive behavior of North American species, *Pungitius pungitius*, the nine-spined stickleback; or *Eucalia inconstans*, the Brook stickleback, for comparison with *Gasterosteus aculeatus*. It would be of further value to see if the same sign stimuli determined for the European three-spined species would release similar behavioral responses in North American species. Such a study should give a substantial introduction to literature search and opportunity for observational and experimental research.

Problem

The basic problems involve describing the reproductive behavior of a North American species of stickleback and also comparing the effects of sign stimuli on this species with information on the European three-spined species. This problem involves five stages: (1) review of literature on stickleback behavior and sign stimuli, (2) establishment of a successful aquarium, (3) observational study of reproductive behavior, (4) experimental study of sign stimuli, and (5) analysis of data and preparation of a report.

Suggested Approach

Keep detailed records throughout the establishment of the aquarium and the observational and experimental research. Measure and record major factors of the environment such as light and water temperature. Try regulating them experimentally. Still and motion photography could be used for review and documentation. As a final caution, make several trials with several different sets of fish to establish any indication of "average" behavior and its variations.

References · general

1. Scott, J. P. 1958. Animal behavior. University of Chicago Press, Chicago.
2. Tinbergen, N. 1951. The study of instinct. Oxford University Press, New York.

· specific

3. Craig-Bennett, M. A. 1931. The reproductive cycle of the three-spined stickleback, *Gasterosteus aculeatus* Linnaeus. Philos. Trans. Roy. Soc., London, series B, 219:197-279.
4. Greenbank, J. and P. R. Nelson. 1959. Life history of the three-spined stickleback, *Gasterosteus aculeatus* Linnaeus, in Karluk Lake and Bare Lake, Kodiak Island, Alaska. U. S. Fish and Wildlife Service, Fishery Bull. 153(59):537-559.
5. Hubbs, C. L. and K. F. Lagler. 1949. Fishes of the Great Lakes Region. Cranbrook Institute of Science. Bulletin No. 26, Bloomfield Hills, Michigan.
6. Klopfer, P. H. 1970. Behavioral ecology. Dickenson Pub. Co., Encino, California.
7. Tinbergen, N. 1942. An objective study of innate behavior of animals. Biblioth. Biotheor. 1:39-98.
8. Tinbergen, N. and J. J. A. van Iersel. 1947. 'Displacement reaction' in the three-spined stickleback. Behavior 1:56-63.
9. Vrat, V. 1949. Reproductive behavior and development of the three-spined stickleback, *Gasterosteus aculeatus,* of California. Copeia (4): 252-260.

4 / BEHAVIOR OF PHOTOSYNTHETIC BACTERIA TOWARD LIGHT

Clyde S. Barnhart, Sr.
U. S. Army L. W. L.
Aberdeen Proving Ground, Maryland 21005

Background

Some sulfur bacteria derive energy for the manufacture of their cell components and for their other metabolic activities by using light to bring about reactions involving sulfur or its compounds. This process is called bacterial photosynthesis. Some biologists have speculated that a key to the synthesis of carbohydrates from inorganic compounds may be found in the photosynthetic sulfur bacteria before it is found in green plants.

The striking motion of bacteria may be observed with a compound microscope. The moving bacteria can be made to reverse their direction of travel simply by passing one's hand between the light source and the mirror of the microscope. This reversal of direction on the part of the sulfur bacteria may be brought about again and again by repeating the interception of the light, momentarily darkening the slide.

Problem

Much can be learned of the nature of this "shadow" sensitivity exhibited by the sulfur bacteria by varying the frequency and duration of the period of interruption of the light. Some questions to be answered are:

1. What is the minimum period of darkness that will elicit a reversal-of-direction response on the part of a sulfur bacterium?

2. How often can this response be repeated? Is there fatigue, or loss of sensitivity, as the darkening of the slide is repeated at close intervals of time?

3. What is the variation within the colony of the minimum period of darkenss required to elicit reversal of direction?

4. What is the characteristic pattern for this response in different species of photosynthetic bacteria?

5. What are the effects of temperature, population density, light intensity, and monochromatic light on this response?

6. Is there an effect that can be related to colony vigor?

7. What effects do antibiotics have on this behavior?

8. What is the effect of intermittent light such as 60-cycle fluorescent or light flashes of greater or lesser frequency? (A Strobotac is a convenient light source for this experiment.)

More difficult questions would be:

1. What is the time lag in the reversal-of-direction response?

2. What is the minimum intensity of light that will produce a reversal of direction?

3. Do sulfur bacteria respond to sudden lowering in intensity of light in the same way that they do to light interceptions? If so, what is the least light change that will bring about a response?

4. If sulfur bacteria become fatigued in their response to light change, what is the recovery rate?

Other questions will suggest themselves as the experiment proceeds.

Sulfur bacteria can be found swimming in quiet water next to the mud in places where sulfurous substances decompose. Likely places are near the discharge of a household septic tank, sewage plant, or meat packing plant. A teaspoonful or so of the mud in a glass jar half filled with the water should provide enough bacteria for study. Be sure to keep the jar in a well-lighted window. Adding a pinch of sulfur to the jar should further enrich the media and favor the growth of the sulfur-utilizing bacteria.

Suggested Approach

The equipment for this study need not be elaborate. As a beginning, a compound microscope, slides, cover slips, medicine dropper, cultures of sulfur bacteria, a light source, and a device for the physical interruption of the light are sufficient.

The light-interrupting device could be electronic, or it could be a camera shutter with the springing action reversed in such a way that the shutter springs shut for measured intervals of time such as one second, 1/10 second, 1/50 second, 1/100 second, 1/200 second, and so forth. A simple pendulum, falling by gravity, placed between the light and the mirror, might well be used. Wooden dowel rods of varying diameters and/or lengths (distance from pivot to light beam) could be used as pendulum devices for varying and controlling the period of interruption of the light.

A spoked wheel rotated at different controlled rates of speed could also be used as a device for rapidly repeating light interruptions.

Other means will suggest themselves as the experiment proceeds, and other means of controlling frequency and duration of light interruptions may be devised.

References · general

1. Bracken, A. 1955. The chemistry of microorganisms. Pitman Publishing Corp., New York.
2. Leathen, W. W. 1961. Sulphur bacteria in the encyclopedia of the biological sciences. Peter Gray, ed. Reinhold Publishing Corporation, New York.
3. McElroy, W. D. and B. Glass, eds. 1961. A symposium on light and life. The Johns Hopkins Press, Baltimore.
4. Salle, A. J. 1961. Fundamental principles of bacteriology. 5th ed. McGraw-Hill Book Co., New York.
5. Veen, R. van der and G. Meijer. 1960. Light and plant growth. The Macmillan Co., New York.

· specific

6. Gest, Howard, Anthony San Pietro, and L. P. Vernon, eds. 1963. Bacterial photosynthesis: a symposium sponsored by the Charles F. Kettering Research Laboratory. The Antioch Press, Yellow Springs, Ohio.
7. Gunsalus, I. C. and R. Y. Stanier, eds. 1962. Bacteria: a treatise on structure and function. Biosynthesis 3. Academic Press, New York.
8. Society of American Bacteriologists. 1957. Manual of microbiological methods. McGraw-Hill Book Co., New York.

5/ *DILEPTUS*– A PROTOZOAN PREDATOR UPON METAZOAN ANIMALS

Harley P. Brown
Department of Zoology
University of Oklahoma
Norman, Oklahoma 73069

Background

Among the commonly encountered protozoa, many are predatory. Of the freshwater forms, amoebas and testaceans, such flagellates as *Peranema*, and a variety of ciliates including *Didinium, Coleps, Bursaria*, and *Dileptus*, are familiar "carnivores." *See* Calkins (1), Jahn and Jahn (2), Kudo (3). In general, these predators swallow their victims whole. Thus, an obvious restriction is placed upon the nature of the prey: victims must be small enough to be engulfed by the predator. It is not surprising, then, that few metazoan organisms are included on the menu, since only such metazoa as rotifers, gastrotrichs, and a few other midgets are within the vulnerable size range. *Dileptus*, however, is a striking exception in its mode of feeding. As it creeps or swims about, its proboscis incessantly waves to and fro or probes crevices and objects encountered. If the ventral surface of the proboscis touches some hapless ciliate—even a giant *Stentor* many times larger than the *Dileptus*—the portion of the victim which was touched disintegrates immediately. The *Dileptus* then proceeds to engulf the disintegrated material. In the case of small ciliates, the entire cytolyzed victim is engulfed; in the case of proportionately huge victims such as *Stentor*, only the cytolyzed portion is engulfed—although this may be of greater volume than the *Dileptus* itself. *See* Dragesco(5), Dragesco and Metain(6), Hayes(7), Jones and Beers(9), Visscher(10).

Possessing an efficient mechanism for feeding upon animals much larger than itself, *Dileptus* is potentially predatory upon a variety of metazoa. This ciliate has been demonstrated to be capable of destroying even such organisms as hydras and planarians, which are themselves quite gluttonous predators. *See* Brown and Jenkins (4). In fact, *Dileptus* was found to have wiped out a laboratory culture of thousands of baby planarians. An interesting further development is that *Dileptus* which have been fed exclusively upon planarian tissues commonly develop into monsters of abnormal size and form. *See* Janovy (8).

Suggested Problems

1. What factors render some kinds of animals vulnerable to attack by *Dileptus,* whereas other animals are relatively safe? (The presence of a cuticle, the secretion of mucus, and certain behavioral patterns are probably important.)

2. What is the nature of the "toxic trichocysts" of *Dileptus* by which the victim is cytolyzed? (It is capable of causing cell membranes to dissolve almost instantly. This should provide a clue to the chemical nature of the substance or substances involved.)

3. What prevents one individual *Dileptus* from injuring or destroying another? Does it fail to discharge the toxic trichocysts, or is the pellicle unaffected by the toxin?

4. What is there about a diet of planarian tissues that induces monster formation in *Dileptus*?

   a. Are essential nutrients lacking? Does monsterism develop if other foods supplement the diet? Does a restricted diet of a different food source produce monsterism?

   b. Are monster-inducing substances present? If the latter, are these substances carcinogenic? Do they have similar effects upon other organisms?

Special Considerations

Dileptus is commonly found in ponds and sluggish streams, usually in close association with the substrate. Once a few specimens have been found and transferred to suitable culture vessels, they can be maintained and cultured by the addition of food organisms as needed. (Cultures of *Dileptus* are also obtainable from Carolina Biological Supply Co., Burlington, North Carolina.) Small ciliates such as *Colpidium* or *Tetrahymena* can readily be cultured separately and added to the *Dileptus* cultures as food at appropriate intervals. In many cities, the tap water may be detrimental or lethal to cultures. A source of pond or lake water may be necessary. As with most predators, it is difficult or impossible to achieve great population density such as we find among paramecia. Good cultures require considerable care and attention.

A good stereoscopic dissecting microscope with magnifications ranging from 10X to 30X or 50X (for example, a Bausch & Lomb Stereo Zoom binocular microscope) and a good spotlight with a rheostat (for example, an American Optical Universal Lamp) are the most useful instruments with which to work, although a compound microscope should be available for occasional use. Culture vessels in which the specimens may be observed include microdiffusion dishes, tissue culture dishes, and small petri dishes. Plastic vessels are satisfactory. If small vessels are used, care must be taken to prevent undue loss of water (for example, the small dish may be kept in a larger dish containing a source of moisture).

Problems 1 and 4 in the list are more likely to be feasible than are 2 and 3. Any of the problems will necessitate establishment of cultures and acquisition of experience in handling the specimens.

References · general

1. Calkins, G. N. 1933. The biology of the protozoa. 2d ed. Lea & Febiger, Philadelphia.
2. Jahn, T. L., and F. F. Jahn. 1949. How to know the protozoa. William C. Brown Co., Dubuque, Iowa.
3. Kudo, R. 1966. Protozoology. 4th ed. Charles C. Thomas, Springfield, Ill.

· specific

4. Brown, H. P. and M. M. Jenkins. 1962. A protozoan (*Dileptus*; Ciliata) predatory upon metazoa. Science 136(3517):710.
5. Dragesco, J. 1963. Revision du genre *Dileptus*, Dujardin 1871 (Ciliata Holotricha) (Systématique, cytologie, biologie). Bull. Biol. de la France et de la Belgique 97(1):103-145.
6. Dragesco, J. and C. Metain. 1948. La capture des proies chez *Dileptus gigas* (Cilie Holotriche). Bull. Soc. Zool. Fr. 73:62-65.
7. Hayes, M. L. 1938. Cytological studies on *Dileptus anser*. Trans. Am. Microscop. Soc. 57(1):11-25.
8. Janovy, J. 1963. Monsterism in *Dileptus* (Ciliata) fed on planarians (*Dugesia tigrina*). J. Protozool. 10(4):428-430.
9. Jones, E. E. and C. D. Beers. 1953. Some observations on structure and behavior in the ciliate *Dileptus monilatus*. J. Elisha Mitchell Sci. Soc. 69(1):42-48.
10. Visscher, J. P. 1923. Feeding reactions in the ciliate *Dileptus gigas*, with special reference to the function of the trichocysts. Biol. Bull. 45:113-143.

6 / RESPONSES OF ALGAE OR PROTOZOA TO ULTRAVIOLET IRRADIATION

H. S. Ducoff
Department of Physiology and Biophysics
University of Illinois
Urbana, Illinois 61801

Background

Readily observed changes in X-irradiated or ultraviolet-irradiated cells include a temporary suppression of cell division in the entire irradiated cell population, and the death of a proportion of the cells. The suppression of division occurs without impairment of synthesizing capacity, so that the cells increase in size, protein content, carbohydrate content, and other features before "recovery" or the resumption of division. This block in division is sometimes referred to as "G_2" or "pre-prophase arrest" because cells in mitosis at the time of exposure usually complete that mitosis, but no new mitoses appear. The death of irradiated cells sometimes occurs with rapid disintegration, or lysis, or with gradual nuclear degeneration (pycnosis). Most characteristically, death occurs at the time the irradiated cell begins division.

There are, in addition, inherited changes brought about by irradiation. These include nonspecific impairment of growth, gene alterations, and chromosomal damage: These are observed in the progeny of these irradiated cells. Furthermore, cytoplasmic organelles such as chloroplasts of *Euglena* (10) may suffer UV damage so that the cells are "bleached," and this cytoplasmic change is passed on to all succeeding cell generations.

The responses to UV radiation of more than 50 species of algae and protozoa (5,9) are known. Most respond in the manner indicated, but there are some notable exceptions. The brackish water alga, *Brachiomonas submarina*, for example, shows no division delay even after supralethal X-ray doses, but does show division delay after moderately low UV doses (2).

The flagellate, *Chilomonas paramecium* (3), and several ciliates (5) undergo two phases of division delay after UV exposure, but only one after X-irradiation. Biologists are well aware of the dictum, "Treasure your exceptions!" Elucidation of patterns of response to radiation in a wider range of species of algae and protozoa would constitute a valuable contribution.

A phenomenon peculiar to UV exposure is that of photoreactivation (7,8), in which illumination with visible (white or blue) light shortly after UV irradiation greatly diminishes the effect of the UV. Less efficient repair processes go on in the dark, and recent work (6,7) emphasizes that the mechanisms for "dark repair" of UV damage are very similar to those for repair of X-ray damage; this lends greater significance to studies of UV effects.

Suggested Approach

The source of irradiation might be a 15-watt germicidal lamp, such as the General Electric G15T8 (1,8), mounted in a fluorescent fixture at a fixed distance—say, 100 cm or more—from the experimental material. The exposures required range from about 15 seconds to several minutes, depending on species and phenomenon studied. Therefore, the greater the distance, the longer the exposure time (the inverse square law operates here) and the lower the error in repeating the dose.

If the organisms studied are large enough so that individuals can be isolated (11) in separate watch glasses, or depressions in glass or plastic plates, the proportion surviving, duration of division block, and growth rate after resumption of division can all be measured. With smaller organisms, survival is difficult to measure unless the individuals give rise to colonies; but division block can be detected and measured by growing mass cultures, decanting samples at various time intervals, fixing the samples with a drop of Lugol's iodine or similar fixative, and counting in a hemocytometer or other chamber (3).

One might examine other phenomena by choosing appropriate organisms. For example, it might be interesting to measure mortality patterns, and division block following UV irradiation, in one or more species of algae or protozoa not previously studied by radiobiologists. Conceivably, some photosynthetic organisms might be considered. Some appear to divide only at night. Would they show differences in sensitivity if irradiated at different hours of day or night? Some protozoa and slime molds have both flagellate and ameboid forms. Do these differ in radiation sensitivity? Is the conversion or differentiation from one form to the other affected by radiation? (Sublethal doses of X rays delay pupation in insect larvae.) What differences are found in the radiation response of sexual and asexual stages in organisms such as *Vorticella* or *Oedogonium*? Is fruiting in a mold accelerated or delayed when vegetative growth is inhibited by radiation? Another project might consist of comparing division block and/or lethality in several species, or in studying a number of forms, or specialized phenomena, in some one uniquely suitable species.

Some Precautions and Technical Tips

Ultraviolet irradiation is not penetrating, and therefore not very hazardous. But the cornea of the eye is very sensitive to UV, and glass or plastic goggles should be worn at all times when the germicidal lamp is used. Ordinary eyeglasses do not offer complete protection, because rays may be reflected around the lenses. Care should also be exercised that bare skin, particularly of the face, should not be exposed too close to the lamp, or exposed for any extended period of time.

In order to avoid photoreactivation, it is best to irradiate in a dimly lit room, preferably one illuminated with only red or yellow light, and to keep material in the dark, or in red or amber flasks or bottles, for at least a few hours after UV exposure. This may complicate the experimental design when photosynthetic organisms are involved but, again, gold fluorescent light (8), yellow light, or bright red light will support photosynthesis without supplying any of the wavelengths necessary for photoreactivation of UV damage.

Every experiment, of course, should include unirradiated controls; in addition, it is desirable to include one or more groups treated exactly the same as groups in previous experiments. To insure uniformity of exposure, it is helpful to mount the irradiation vials on a phonograph turntable, which is kept revolving slowly during the irradiation. It is also useful to have a shutter mechanism to time exposures accurately.

References

1. Buttolph, L. J. 1956. Practical applications and sources of ultraviolet energy. In A. Hollaender, ed., Radiation biology, Vol. 2, p. 41, McGraw-Hill Book Co., New York.

2. Ducoff, H. S. and B. D. Butler. 1965. Photoreactivation of ultra-violet-induced suppression of division in *Chilomonas paramecium*. Exptl. Cell Res. 40:104.

3. _____, and E. J. Geffon. 1965. The effect of radiations on replication in *Brachimonus submarina* Bohlin. Radiation Research 24:563.

4. Giese, A. C. 1953. Protozoa in photobiological research. Physiol. Zool. 26:1.

5. _____. 1967. Effects of radiation upon protozoa. In T.-T. Chen, ed., Research in protozoology, Vol. 2, p. 276, Pergamon Press, Inc., New York.

6. Hanawalt, P. C. 1969. Radiation damage and repair *in vivo*. In C. P. Swanson, ed., An introduction to photobiology, p. 53, Prentice-Hall, Inc., Englewood Cliffs, N. J.

7. _____ and R. H. Haynes. 1967. The repair of DNA. Sci. Amer. 216 (2):36.

8. Jagger, J. 1967. Introduction to research in ultraviolet photobiology. Prentice-Hall, Inc., Englewood Cliffs, N. J.

9. Kimball, R. F. 1956. The effects of radiation on protozoa and the eggs of invertebrates other than insects. In A. Hollaender, ed., Radiation biology, Vol. 2, p. 285, McGraw-Hill Book Co., New York.

10. Lyman, H., H. T. Epstein, and J. A. Schiff. 1961. Studies of chloroplast development in *Euglena*. I. Inactivation of green colony formation by UV light. Biochem. Biophys. Acta 50:301.

11. Sonneborn, T. M. 1950. Methods in the general biology and genetics of *Paramecium aurelia*. J. Exptl. Zool. 113:87.

7 / THE INFLUENCE OF CERTAIN DRUGS ON THE BEHAVIOR OF PROTISTS

Background

Helene N. Guttman
College of Liberal Arts and Sciences
Department of Biological Sciences
Box 4348, Chicago, Illinois 60680

In higher animals, sensitivity to stimuli and the responses to them are mediated by the nervous system. Although irritability can be altered by a wide variety of drugs, the exact nature of the drug action is difficult to determine. One way of attacking this problem is to study the effect of a drug on a unicellular organism or protist. We generally classify the protist *Euglena* as a plant because it can convert light energy to chemical energy (photosynthesis). An animal protist such as *Tetrahymena* lacks this power, but it can consume and metabolize particles. Although they lack a nervous system they do demonstrate irritability and response. A change in response caused by a drug may yield a clue to the action of drugs that increase or decrease response in higher animals. It might also be possible to study the chemical changes in the protoplasm brought about by these drugs.

Problems

1. To analyze the effects of certain drugs on the behavior of motile protists such as *Euglena* and *Tetrahymena*.

2. To compare the effects of drugs on these two forms and to relate the findings to effects of the same drugs on higher plants and animals.

Suggested Approach

Obtain a supply of *Tetrahymena* and *Euglena* from the American Type Culture Collection, 2112 M Street, N.W., Washington, D. C. 20037. You can maintain your own cultures by preparing sterile medium composed of: cane sugar 1.0%; Trypticase (hydrolyzed casein) 0.5%; yeast extract 0.1%. Add tap water to 100 ml and adjust to pH 6.5. Then you may transfer your cultures from an old to a new tube of medium in order to build up a large population. It is necessary to transfer with sterile pipettes and to use aseptic techniques throughout or your cultures will soon become contaminated. *Tetrahymena* is ciliated, while *Euglena* is flagellated. Their speed of movement can be studied by microscopic observation of a suspension of the organisms on a glass depression slide. The student must learn to recognize "normal" motility. The effect of drugs on this aspect of behavior can then be tested by placing a sample of the *Tetrahymena* culture in a glass well to which has been added a small amount of the test substance. (It is not necessary to observe aseptic techniques for the test.) For the first test series, use an antihistamine known to produce drowsiness in higher animals. Test drugs may be obtained through the help of a local physician or pharmacist. Decreased motility may be considered the "protist equivalent" of drowsiness. The study can be extended to other sedatives and to stimulants. A final step would be to study chemical changes in the composition of the motile and nonmotile *Tetrahymena* by paper chromatographic analysis.

The particular response of the *Euglena* to light provides still another analytical approach. These plant cells will mass behind the colors that represent the action spectra of their photosensitive pigments. These colors can be determined by shielding the *Euglena* from all light except that filtered through colored cellophane and then observing the areas in which the organisms congregate. When the colors to which the *Euglena* responds are determined, the time taken for massing should be measured ($1\frac{1}{2}$ to several hours). The influence of a drug on both the time taken for massing and the colors to which the *Euglena* responds can be determined by comparing the rate of massing and the number of organisms in a particular color area when the organisms are subjected to the drug with the control.

For more accurate work, it is most desirable to use monochromatic light, or at least to narrow the band of a given color to a relatively small range of wavelengths. Colored cellophanes give a misleading impression from this point of view. For example, blue cellophane also transmits red light; green cellophane transmits both red and blue; and red cellophane also transmits blue. Emulsions used in theatrical work are inexpensive and superior to cellophane with respect to purity of color.

One way of delimiting the range of wavelengths is to use certain specific combinations of light sources and filters. Several useful combinations are given below, together with the wavelengths that are transmitted.

Fluorescent Light Tube Color	No.	Cinemoid* Filter Color	Wavelengths Transmitted	Peak Trans- mission	Color of Light
Blue	19	Dark Blue	375-500 mu	420 mu	Blue
Green	24	Dark Green	475-575 mu	525 mu	Green
Gold	26	Mauve	575-700 mu	650 mu	Red

Additional Problems

It would also be interesting to determine whether compounds such as streptomycin and antihistamines, or compounds that screen out certain colors, affect the ability of *Euglena* to seek out and move toward specific wavelengths of light.

Additional experiments could be designed to find compounds that reverse the action of the drugs that were found to affect behavior of either *Tetrahymena* or *Euglena*.

*Cinemoid filters (size 20 x 24 inches) may be obtained from Kliegl Brothers Co., 321 W. 50 Street, New York, N. Y. 10019.

References · general

1. Block, R. J., E. L. Durrum, and G. A. Zweig. 1958. A manual of paper chromatography and paper electrophoresis. 2d ed. Academic Press, New York.
2. Corliss, J. O. 1961. The ciliated protozoa. Pergamon Press, New York.
3. Lwoff, A. and S. H. Hutner, eds. 1951-1963. Biochemistry and physiology of protozoa. 3 vols. Academic Press, New York.
4. Wolken, J. J. 1961. *Euglena*. Rutgers University Press, New Brunswick, N. J.

· specific

5. Halldal, P. 1958. Action spectra of phototaxis and related problems in Volvocales, Ulva-Gametes and Dinophyceae. Physiol. Plantarum 11:118.

6. _____. 1959. Factors affecting light response in phototactic algae. Physiol. Plantarum 12:742.
7. Nathan, H. A. and W. Friedman. 1962. Chlorpromazine affects permeability of resting cells of Tetrahymena pyriformis. Science 135: (3506):793-794.
8. Sanders, M. and H. A. Nathan. 1959. Protozoa as pharmacological tools: the antihistamines. J. Gen. Microbiol. 21:264.

8/ ON THE BEHAVIOR OF HYDRA

Howard M. Lenhoff
School of Biological Sciences
University of California
Irvine, California 92664

The simple freshwater hydra appears ideal for investigating some metazoan behavioral processes. These animals respond in a specific manner to a relatively few external stimuli, while they have none of the complexities usually associated with organisms possessing a highly developed nervous system. And, most important for you, it is practically feasible to study the behavior of hydra with a minimum of equipment.

The behavior of these animals is remarkably geared for food capture; three receptor systems assist them in securing food. There is a phototactic one, which may lead them to areas where food is abundant; a chemotactile one to help capture prey; and a chemical receptor, to coordinate the movements involved in feeding.

PHOTOTAXIS

Background

In 1744 Abraham Trembley of Geneva reported some remarkable and careful experiments using hydra as the experimental animal. Among these were some showing that hydra move toward light. That discovery is even more remarkable since hydra have not yet been shown to have any structures that are considered light receptors.

Suggested Problems and Approaches

First, find a method for measuring the movements of hydra toward light; it is important to express your results as numbers. For example, you may want to measure the time it takes hydra to move toward the light. Once you have worked out a method, then be certain your experimental system is foolproof. That is, are you taking adequate measures to be certain that the light is not heating the water? Heat can produce a temperature gradient which may affect the movement of the animal. Does your container produce reflections which may cause the hydra to move in other directions?

Once your system is perfected, you may want to ask questions such as: Where are the light receptors located? Will dissected parts of the hydra (head or body) move independently toward light? Faster? Are the tentacles required for phototaxis? Are hydra attracted to different wavelengths of light? Do hydra of different color behave the same toward light? (You can color hydra black, for instance, by feeding them cut-up tadpoles; or red, by feeding them clotted blood). Does the physiological state of the hydra affect their movement toward light? Will a young bud behave the same as a budding parent? Will starved hydra move faster or slower than fed animals? Can hydra adapt to light and no longer respond? Will drugs interfere with the response to light? Can light cause negative phototaxis?

CHEMOTACTILE RESPONSE

Background

The most notorious structures of hydra are the stinging capsules, also called nematocysts, or cnidae. As implied in the name, all members of the phylum Cnidaria possess cnidae. These structures, which are basically coiled harpoons, or lassoes, are triggered to explode when bumped into by a small swimming organism, such as a *Daphnia*, resulting in the capture of the prey. During locomotion such as phototaxis, the hydra uses some nematocysts to attach itself to the substratum. Since each nematocyst acts independently of others, and does not appear to be under the control of any central coordinating system, they are called "independent effectors." Despite all that is known about these structures, we still know very little regarding factors affecting their discharge.

Suggested Problems and Approaches

Just as in the phototaxis experiments, you need to devise an assay for the study of the discharge of nematocysts. If a microscope is available, you might want to count the number of nematocysts that are discharged from hydra tentacles. Since nematocysts remain adhered to the prey after feeding, you might count the loss of nematocysts from the tentacle. Another way to assay for the functioning of nematocysts is to attempt to feed hydra prey (*Artemia* larvae) under different experimental conditions. If the hydra captures the prey, then the nematocysts must have discharged. You may want to isolate the nematocysts, or flatten the tentacle on a slide, and then measure the number of nematocysts discharged before and after treatment. Can you think of another way?

You can now ask: How many nematocysts discharge "spontaneously" under the assay conditions? How many discharge during feeding? Can food extracts cause nematocysts to explode? Try poking the tentacle with a glass rod—and then with a glass rod that has been immersed in various solutions such as food extracts, egg white, blood, etc. Will pretreating the food (like boiling the *Artemia*) alter its ability to discharge nematocysts? Can this ability to discharge, once lost, be restored? Are the various ions present in your test solution required for discharge? Try modifying the calcium content or adding magnesium. Add copper, poisons, metals, etc. What is the effect on discharge of drugs, alcohol, or aspirin? Vary pH, temperature. Consider the physiological state of the hydra. Will nematocysts discharge immediately after the hydra have eaten? If not, how soon after?

CHEMICAL COORDINATION OF FEEDING MOVEMENTS

Background

Although hydra are considered to have a very simple nervous system (some scientists even doubt that hydra have nerve cells), these animals will swallow only live or recently killed prey, not inanimate or dead objects. The mystery of the control of hydra's feeding reaction was recently solved by W. F. Loomis. He demonstrated that when the prey is pierced by the harpoonlike nematocysts, body fluids ooze from the wounds. In these fluids (and in fluids of virtually all living organisms) is the ubiquitous tripeptide, reduced glutathione, a compound which Loomis shows can make some species of hydra eat. Even low concentrations of glutathione can make hydra eat nearly any inanimate object offered it.

Suggested Problems and Approaches

Again, you need a quantitative assay. One successful approach has been to measure the time during which the mouth stays open in the presence of the feeding activator. To do this you need a watch with a second hand and either a dissecting microscope or a 5 to 10 power mounted hand lens. Before beginning your quantitative experiments, spend some time observing hydra eating live prey in order to become familiar with their normal feeding movements. Thus you will not confuse other tentacle and mouth movements with feeding.

What kinds of animal can hydra eat? Can they eat any plants? Dried yeast? From these observations can you make any deductions about the feeding behavior of hydra in nature? Will hydra respond to extracts made from their prey, or from some animals they do not ordinarily eat? What is the smallest dilution of extract that induces feeding? Can you treat active extracts to make them inactive? Will this tell you something about the activator? Using methods like chromatography, can you isolate the feeding activator from the extracts? Can other chemicals from your laboratory activate feeding? Will pure reduced glutathione activate feeding of your hydra? (Reduced glutathione can be purchased from Sigma Chemical Co., St. Louis, Mo.) What environmental factors, such as pH and temperature, affect the induction of feeding by food extracts? What are the effects of ions such as calcium, magnesium, sodium, potassium, chloride, nitrate, etc.? Can mixtures of ions affect feeding differently than does each of the above ions? What is the effect of pH? Of drugs? Do the same compounds that affect the feeding response also affect nematocyst discharge?

Now consider your hydra. Can they adapt to the activator? Can they be poisoned? How do starved hydra respond compared to fed hydra? Does the age of the animal make a difference? The species? The number of tentacles present (try removing some)? If you live by the seashore, try other coelenterates such as jellyfish, sea anemones, or colonial hydroids. Why not try animals from other phyla? For example, will food extracts make planaria extend the proboscis? If this happens, try some of the experiments suggested for hydra.

Possible Pitfalls

Before beginning any of the suggested experiments, it is essential that you be able to maintain hydra in your laboratory under the best possible conditions. Sick ("depressed") hydra will not give consistent or natural responses. You can be assured that your hydra are in prime condition if they grow at a logarithmic rate, doubling every 1½ to 2 days. Furthermore, to insure repeatability of your results, try to control the many possible variables that might affect the response of hydra. For example, control the feeding schedule, the composition of the hydra's medium, and the temperature.

References · general
1. Hyman, L. H. 1940. The invertebrates: protozoa through cteno-phora. McGraw-Hill Book Co., New York.
2. Lenhoff, H. M. and W. F. Loomis, eds. 1961. The biology of hydra and of some other coelenterates. University of Miami Press, Coral Gables, Fla.
3. Lenhoff, H. M., L. Muscatine, and L. V. Davis. 1971. Experimental coelenterate biology. University of Hawaii Press, Honolulu.
4. Prosser, C. L. and F. A. Brown, Jr. 1961. Comparative animal physi-ology. 2d. ed. W. B. Saunders Co., Philadelphia.

· specific
5. Fulton, C. 1963. Proline control of the feeding reaction of *Cordylo-phora*. J. Gen. Physiol. 46:823-837.
6. Lenhoff, H. M. 1968. Behavior, hormones, and hydra. Science 161 (3840):434-442.
7. _____ and R. D. Brown. 1970. Mass culture of hydra: an improved method and its application to other aquatic invertebrates. Laboratory Animals 4:139-154.
8. Lindstedt, K. J., L. Muscatine, and H. M. Lenhoff. 1969. Valine acti-vation of feeding in the sea anemone *Boloceroides*. Comp. Biochem. Physiol. 26:567-572.
9. Loomis, W. F. 1953. The cultivation of hydra under controlled con-ditions. Science 117:565-566.
10. _____. 1955. Flutathione control of the specific feeding reactions of hydra. Ann. N. Y. Acad. Sci. 62:209-227.
11. _____ and H. M. Lenhoff. 1956. Growth and sexual differentiation of hydra in mass culture. J. Exptl. Zool. 132:555-574.
12. Yanagita, T. M. 1960. Physiological mechanisms of nematocyst re-sponses in sea anemone. Comp. Biochem. and Physiol. 1:140-154.

9/ THE AMINO ACID CONTENT AND REGENERATIVE PROPERTIES OF PLANARIA *DUGESIA DOROTOCEPHALA*

Sister M. C. Lockett
Director of Department of Biology
University of Dallas
Dallas, Texas 75201

Background

The phenomenon of regeneration is both perplexing and important to biologists. Regeneration, the restoration of lost or amputated parts, is a capacity manifested to a high degree among invertebrates and many cold-blooded vertebrates. It diminishes as we ascend the evolutionary scale, to wound-healing processes in the warm-blooded animals. Planarians, in particular, have been the subject of extensive research. Certain general facts ascertained from this research apply to all the lower animals. In the first place, any piece of such animals usually retains the same polarity it had while in the whole animal. That is, the regenerated head grows out of the cut piece that faced the anterior end in the whole animal, and the regenerated tail grows out of the cut end that faced the posterior end. Another generalization drawn from these experiments is that the capacity for generation is greatest near the anterior end and decreases toward the posterior end.

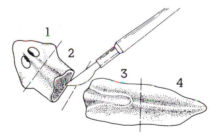

Problem

If the freshwater planarian, *Dugesia dorotocephala*, is cut into four pieces (*see* accompanying figure), Section 3 will produce a large number of abnormal regenerates, some with reduced heads and some completely headless. Sections 2 and 4 produce regenerates with normal heads. A primary purpose of this project is to investigate the possibility of a relationship between amino acid content and the regenerative properties of different sections. Since it has been observed that Section 3 produces abnormalities in head development, while Section 2 and Section 4 produce regenerates with normal heads—Section 2, however, regenerating faster than Section 4—it is our problem to determine:

1. Whether the abnormality in head development of Section 3 may be due to a deficiency of a particular amino acid in this section.

2. Whether there is a greater concentration of a particular amino acid in Section 2 which gives this section the capacity to regenerate faster than Section 4.

Suggested Approach

1. Obtain planaria, *Dugesia dorotocephala*, and keep in enameled (glass) containers filled with dechlorinated tap water. Feed beef liver approximately twice a week; keep in temperature ranging from $16°$ to $20°$ C.

2. After three to four weeks of acclimatization to the laboratory, starve the worms for seven to ten days in order to maintain uniformity among the experimental animals.

3. Cut the worms into four sections as shown in the previous figure and keep sections in dechlorinated water; observe the regeneration of each section—the time it takes and the number of normal head regenerates in each section. At the same time, section worms for the determination of the amino acid content.

4. Determine the amino acid content by paper chromatography.
    a. Hydrolysis
        (1) Place a sample consisting of similar sections of ten planaria in 4 ml of Krebs-Ringer Buffer.
        (2) Add 40 mg of pancreatin to each flask.
        (3) Adjust pH to 8.5 with 0.1 N NaOH (about six drops) and cover with toluene.
        (4) Hydrolyze at $37°$ C for approximately 24 hours.
    b. Extraction
        (1) Acidify hydrolyzate to pH with 5 N HCl (approximately 4 drops). Do not remove toluene.

(2) Add 6 ml Butanol saturated with 0.1 N HCl, shake, centrifuge and decant with eyedropper after 15 minutes.

(3) Add 3 ml Butanol and shake. Centrifuge for 15 minutes and decant. (Add top layer to above decantate.)

(4) Add 3 ml Butanol and shake. Centrifuge for 15 minutes and decant.

(5) Centrifuge the three combined portions and decant with eyedropper to water layer.

(6) Take down to dryness under reduced pressure.

(7) Add 0.3 ml pure Butanol.

(8) Put extract on chromatographic paper.

c. Preparing strips for separation of the amino acids. (The procedure listed below should be run on solutions of pure amino acids simultaneously with the experimental solutions.)

(1) Place 10 or 25 lambda of solution on chromatographic paper 8 cm from bottom or as suggested in Lederer (p. 129).

(2) Place two strips of each sample in each of the two solvents: Butanol-Dioxane-Ammonia (BDN) and Butanol-Acetic Acid-Water (BH). *See* Lederer (p. 308).

(3) Keep cabs in room of constant temperature.

(4) Remove strips from cabs when the solvent front is near the top (time depends upon the length of the strip and the height of the cab).

(5) Hang to dry. When dry, spray with ninhydrin.

(6) Determine R_f values of the experimental strips as well as those of the standard solutions of amino acids. *See* Lederer (p. 138) for determination of R_f value and page 139 for quantitative determination.

Preparation of Solutions

Butanol Dioxane-Ammonia

800 mls normal Butanol

200 mls Dioxane (freshly distilled)

1000 mls 2N NH_4OH—132 mls concentrated to a liter. Shake in separatory funnel. Use upper layer—usually takes overnight for clear separation between layers.

Butanol-Acetic Acid-Water
780 mls normal Butanol
50 mls Glacial Acetic Acid
170 mls distilled water
Shake
Will form one phase
Krebs-Ringer Buffer

27.8	g NaCl
16.89	g Na_2HPO_4
3717.5	ml H_2O
120.0	ml 1.15% KCl
30.0	ml 3.82% $Mg_2SO_4 \cdot 7H_2O$
0.986 ml cone HCl	

Additional Experiments

Try using single amino acids on Section 3 to see whether the added nutrient affects the regenerative properties of these sections. Are there chemical factors contained in the other sections but lacking in Section 3? Prepare homogenates and extracts of each of the other sections and see how they affect the regeneration of Section 3.

How do physical agents such as ultraviolet light affect regenerative properties of planaria? How do chemical agents such as nitrogen mustard, colchicine, maelic acid hydrazide, and other growth-stunting substances affect the regenerative properties of planaria?

References · general

1. Block, R. J., E. L. Durrum, and G. Zweig. 1958. A manual of paper chromatography and paper electrophoresis. 2d ed. Academic Press, New York.
2. Child, C. M. 1941. Patterns and problems of development. University of Chicago Press, Chicago.
3. Hyman, L. H. 1951. The invertebrates. Vol. 2. McGraw-Hill Book Co., New York.
4. Lederer, E. and M. Lederer. 1957. Chromatography. 2d ed. D. Van Nostrand Co., Princeton, N. J.
5. Needham, A. F. 1952. Regeneration and wound-healing. John Wiley & Sons, New York.

6. Simmonds, S. and J. S. Fruton. 1958. General biochemistry. 2d ed. John Wiley & Sons, New York.
7. Thompson, D. W. 1952. On growth and form. 2 vols. 2d ed. Cambridge University Press, New York.

· specific

8. Brøndsted, Agnes and H. V. Brøndsted. 1953. The acceleration of regeneration in starved planarians by ribonucleic acid. J. Embryol. Expt. Morphol. 1(1):49-54.
9. Coward, S. J. 1968. The relation of surface and volume to physiological gradients in planaria. Develop. Biol. 18(6):590-601.
10. _____, F. M. Hirsh, and J. H. Taylor. 1970. Thymidine kinase activity during regeneration in the planarian Dugesia dorothcephala. J. Exp. Zool. 173(3):269-279.
11. Crawford, F. T., F. J. King, and Milton Mogas. 1967. Amino acid analysis of intact planarians by paper chromatography. J. Biol. Psychol. 9(2):34.
12. Flichinger, Reed A. 1959. A gradient of protein synthesis in planaria and reversal of axial polarity of regenerates. Growth 23(3):251-271.
13. Hammett, F. S. 1943. The role of the amino acids and nucleic acid components in developmental growth. Growth 7:331-399.
14. Henderson, R. F. and R. E. Eakin. 1959. Alteration of regeneration in planaria treated with lipoic acid. J. Exptl. Zool. 141(1):175-190.
15. Jenkins, M. M. 1958. The effects of thiourea and some related compounds on regeneration in planarians. Biol. Bull. 116(1):106-114.
16. Kido, T. and Y. Kishida. 1969. Problems on planarian regeneration. Zool. Mag. 77(7):199-206.
17. Morrill, John B., Jr. 1958. A study of amino acids and proteins in two species of Tubularia and their relationship to regeneration. Dissertation Abstrs. 18(5):1906-1907. Florida State University, Tallahassee.
18. Navarra, J. G. and T. Gerne. 1962. Paper chromatography. Duquesne Sci. Counselor 25(3):66-69. (This article has very practical suggestions for separation of amino acids by paper chromatography.)
19. Pedersen, K. J. 1958. Morphogenetic activities during planarian regeneration as influenced by triethylene melanine. J. Embryol. Exptl. Morphol. 6(2):308-334.

33

10/ THE INFLUENCE OF HOST HORMONES ON PARASITE BURDEN

Lewis E. Peters
Northern Michigan University
Marquette, Michigan 49855

Background

Certain recent studies suggest that the sex of a host affects the ability of a parasitic species to become established and to mature in the host. On the basis of these laboratory studies it is inferred that such differences may be attributed to the sex hormones of the host. There has been little or no general survey work to indicate whether significant differences in infection occur between male and female hosts in nature. If such could be determined from field surveys, evidence for that hypothesized relationship would be strengthened. If such differences were found to occur seasonally (at times of increased gonadal activity preceding or accompanying the spawning or mating period) in the case of fish or frog species or other hosts with seasonally limited gonadal activity, such findings would add further support for the hypothesis.

Suggested Approach

Determine a host species that can be studied easily, considering especially (1) its availability in moderate numbers at all times of the year and (2) its parasitic fauna. In respect to the latter, the host should be rather highly parasitized by at least a few different species, preferably of different groups. For example, in one area I found the channel catfish, *Ictalurus punctatus,* rather heavily infected with several species of flukes as well as with at least one species each of tapeworm, throny-headed worm, and roundworm. The hosts should be collected uniformly throughout the year from the same area and preferably be close to the same age, in fish roughly estimated by length and weight or more accurately, perhaps, by scale year counts. If a sufficient number of male and female hosts are examined each month for a year or more, variations associated with sex of the host and/or with activity of the gonads should be easily detected. If time does not permit immediate examination of all hosts collected at one time, the excess can be preserved in formalin (for fish or frogs) or by deep freezing (for birds or mammals). If the digestive tract only is to be examined, it can be removed fresh from the host and preserved. Verification of the scientific names of the parasites by an expert is usually advisable; the parasitologist at a nearby college or university can suggest the expert for a particular group.

An advantage of the proposed investigation is that it requires little space or equipment, mainly a data notebook, collecting and dissecting gear, bottles, dissecting microscope, and compound microscope; nevertheless, the physiologic implications could be most interesting. If the survey indicated a positive relationship between sex and gonadal activity and the parasitic fauna, the more adept student may wish to design and perform experiments to test his hypothesis, using the standard endocrinologic techniques of extirpation of the gonads or hormone injections.

Interpretation of the data would require a basic understanding of elementary statistical methods.

The reference by Stunkard is of special value on account of its bibliography and the book by Cable for its discussion of techniques.

References · general

1. Bailey, N. T. J. 1959. Statistical methods in biology. John Wiley & Sons, New York.
2. Cable, R. M. 1959. An illustrated laboratory manual of parasitology. Burgess Publishing Co., Minneapolis, Minn.

· specific

3. Beck, J. W. 1952. Effect of gonadectomy and gonadal hormones on singly established *Hymenolepis diminuta* in rats. Exptl. Parasitol. 1:109-117.
4. Bennison, B. E. and G. R. Coatney. 1948. The sex of the host as a factor in *Plasmodium gallinaceum* infections in young chicks. Science 107:147-148.
5. Dobson, C. 1961. Certain aspects of the host-parasite relationship of *Nematospiroides dubius* (Baylis). I: Resistance of male and female mice to experimental infections. J. Parasitology 51:173-179.
6. Dudzinski, M. L. and R. Mykytowycz. 1963. Relationship between sex and age of rabbits *Oryctolagus cuniculus* (L.) and infection with nematodes *Trichostrongylus retortaeformis* and *Graphidium strigosum*. J. Parasitol. 49:55-59.

7. Matthies, A. W., Jr. 1959. Certain aspects of the host-parasite relationship of *Aspiculuris tetraptera*, a mouse pinworm. II: Sex resistance. Exptl. Parasitol. 8:39-45.
8. Stunkard, H. W. 1959. Induced gametogenesis in a monogenetic trematode, *Polystoma stellai* Vigueras, 1955. J. Parasitol. 45:389-394.
9. _____. 1970. Trematode parasites of insular and relict vertebrates. J. Parasitology 56(6):1041-1054.

11 / DURATION OF MOLT IN BIRDS AND MAMMALS

Background

J. Murray Speirs
Department of Zoology
University of Toronto
Toronto, Ontario, Canada

The fact that birds molt feathers periodically and that mammals shed hair at intervals is well known. However, the *details* of the molt are known for *very few species.* The length of time required to complete the molt in different regions of the body is poorly documented. For instance, the length of time that the robin *Turdus migratorius* retains its spotted juvenal plumage does not appear to be known with any accuracy.

The time at which molt begins and its duration are influenced by environmental factors, notably by the light regime to which the animal is exposed.

Suggested Approach

It would be best to work with some common, readily available bird or mammal, perhaps regular patrons at feeding stations, perferably resident species available at all seasons. English sparrows, starlings, squirrels, or chipmunks might be suitable subjects for study. English sparrows and starlings usually are not protected by law; there still are research problems remaining for these common and well-known birds. At the national level, advice on the matter of permits for trapping and keeping birds in captivity should be sought from the U. S. Fish and Wildlife Service, Patuxent Wildlife Research Center, Laurel, Maryland 04105. No one should attempt to learn to trap birds without the personal assistance of an expert. For advice on a state or local level one should contact the local state Conservation Department. Local Audubon societies, bird-banding associations, and natural history museums should also be able to give good advice on a local level.

Since molt is affected by length of day, the study animals should not be retained indoors where they would be exposed to more (or less) light than in the natural environment. Caged animals might be used later to demonstrate the effects of the light regime on the incidence and progress and duration of molt, if so desired.

Birds might be given colored leg bands to be sure that the same individuals are being studied throughout the course of the study. If mammals are used they should also be marked in some manner to ensure that the same individuals are being observed.

The student should become familiar with the position and terminology of the feather tracts, then record the date of beginning of molt and its progress and termination in each feather tract. Sketches or photographs would help to illustrate the progress and some graphical method of recording the data should also be devised. The main flight feathers of wing and tail (remiges and rectrices) are usually fixed in number for particular species and the method of numbering these is given in various manuals. The order in which these are molted and replaced should be recorded. In the case of caged individuals these feathers might be mounted in their proper order as molted and the date when each was molted and replaced indicated on the mounting sheet. In the case of mammals, outline diagrams showing the areas occupied by new fur could be drawn at intervals of a few days. Dorsal, ventral, and perhaps lateral views might be used.

References · general

1. Butcher, E. O. 1951. Development of the pilary system and the replacement of hair in mammals. Ann. N. Y. Acad. Sci. 53:508-516.
2. Dwight, J., Jr. 1900. The sequence of plumages and moults of the passerine birds of New York. Ann. N. Y. Acad. Sci. 13:73-360.
3. Pettingill, O. S. 1950. A laboratory and field manual of ornithology. Burgess Publishing Co., Minneapolis, Minn. pp. i-v; 1-248.

· specific

4. Brown, Frank A., Jr. and Marie Rollo. 1940. Light and molt in weaver finches. Auk 57(4):485-498.
5. Constantine, D. G. 1957. Color variation and molt in *Tadarida brasiliensis* and *Myotis velifer*. J. Mammal. 38(4):461-466.
6. Gottschang, J. L. 1956. Juvenile molt in *Peromyscus leucopus noveboracensis*. J. Mammal. 37(4):516-520.
7. Gunn, C. K. 1932. Color and primeness in variable mammals. Am. Naturalist 66:546-559.

8. Hadwen, S. 1929. Color changes in *Lepus americanus* and other animals. Can. J. Res. 1(2):189-200.

9. Hart, J. S. 1956. Seasonal changes in insulation of the fur. Can. J. Zool. 34(1):53-57.

10. Kabat, C., D. R. Thompson, and F. M. Kozlik. 1950. Pheasant weights and wing molt in relation to reproduction with survival implications. Wisc. Cons. Dept., Tech. Wildlife No. 2:1-26.

11. Lesher, S. W. and S. C. Kendeigh. 1941. Effect of photoperiod on molting of feathers. Wilson Bull. 53(3):169-180.

12. McLaren, I. A. 1958. The biology of the ringed seal (*Phoca hispida* Schreber) in the eastern Canadian Arctic. Fish. Res. Bd. Canada, Bull. No. 118:1-97.

13. Moreau, R. E. 1951. Geographical variation and plumage sequence in *Pogonocichla*. Ibis 93:383-401.

14. _____, A. L. Wilk, and W. Rowan. 1947-1948. The molt and gonad cycle of three species of birds at five degrees south of the equator. Proc. Zool. Soc. London 117: Parts II, III:345-364.

15. Morejohn, G. V. and W. E. Howard. 1956. Molt in the pocket gopher, *Thomomys bottae*. J. Mammal. 37(2):201-213.

16. Munro, J. A. 1941. The grebes. Studies of waterfowl in British Columbia. B. C. Provincial Mus., Occasional Papers, No. 3:1-71.

17. Negus, N. C. 1958. Pelage stages in the cottontail rabbit. J. Mammal. 39(2):246-252.

18. Petrides, G. A. January 1951. Notes on age determination in juvenal European quail. J. Wildl. Mgmt. 15(1):116-117.

19. Rand, R. W. 1956. The Cape fur seal *Arctocephalus pusillus* (Schreber), its general characteristics and molt. Union South Africa, Dept. Comm. & Ind., Div. Fish., Investigational Rept. No. 21:1-52.

20. Sutton, G. M. 1935. The juvenal plumage and post-juvenal molt in several species of Michigan sparrows. Cranbrook Inst. Sci. Bull. 3:1-36.

21. Test, F. H. 1945. Molt in flight feathers of flickers. Condor 47:63-72.

22. Torrey, H. B. and B. Horning. 1925. The effect of thyroid feeding on the molting process and feather structure of the domestic fowl. Biol. Bull. 49(4):275-287.

23. Weller, M. W. 1957. Growth, weights, and plumages of the redhead *Aythya americana*. Wilson Bull. 69(1):5-38.

12 / LENGTH OF INTESTINE AND ENZYME PATTERN IN TADPOLE AND FROG

Background

P. B. van Weel
Department of Zoology
University of Hawaii
Honolulu, Hawaii 96822

Digestive enzymes are produced by digestive glands to accomplish the breakdown of food substances to such a molecular size that they can be resorbed by the intestinal wall and hence utilized by the animal. Since nature is thrifty, it may be expected that the amount of enzymes secreted will fit the amount and kind of food eaten. In general, this seems to be the case, at least with respect to proteases and carbohydrases.

For example, carnivores produce more proteases and fewer carbohydrases than herbivores. However, animals are known to exist which switch diets during their lifetimes. But we know very little about possible changes of enzyme pattern. This problem has recently been studied in the large digestive gland of the African Giant Snail, and here a definite change in the enzyme pattern was found. It would be interesting to explore possible changes in another animal, the frog.

Suggested Problem

Given an animal that shows a definite change in its *normal* diet during its life cycle, (a) does a change in its digestive enzyme pattern occur? (b) if so, *when* does it happen and to what extent?

The larval stage of the frog (the tadpole) is definitely herbivorous, but after metamorphosis the frog is carnivorous. This change puts its stamp on the morphology of the digestive tract. We know that carnivorous animals have a relatively shorter intestinal tract than herbivorous ones. This has long been known to hold for tadpoles and frogs: the former have relatively a much longer digestive canal than the latter. But little is known of the changes with respect to the various stages of development and metamorphosis. Is the change gradual or abrupt? When does it occur in relation to the other changes in metamorphosis? Is there a change in enzyme production? If so, when does this occur? Solving these problems will contribute to an understanding of the physiology of development in these animals.

Suggested Approach

The digestive enzymes are produced by glands in the wall of the digestive tract, as well as by glands outside of the digestive tract connected with it by ducts. In tadpoles the pancreas is small, rather difficult to locate, to excise, and to treat separately from the rest of the intestinal tract. However, because changes in the total amount of enzymes must be determined, an extract of the entire gut with the adhering pancreas can be made, thus appreciably simplifying the technique of extraction. The stomach and esophagus should be extracted separately, because pepsin has such a low pH optimum (pH 2−2.5).

Before extraction, the organs should be weighed quickly and accurately, so that the enzyme strength per unit of tissue weight can be computed. Otherwise, a comparison would not be possible. It must be kept in mind that the physiological state of the animal should always be the same. Hence it is advisable to starve all animals for 24 hours before extraction is undertaken.

To make the extracts, mash up the organs with scissors; add 50-70% glycerol (1 part tissue with 2 parts glycerol is usually adequate), mix well, add a few drops of toluene (to prevent bacterial action), and put into the refrigerator overnight. Then either centrifuge (using the clear supernatant fluid as the extract) or filter through glass wool. To minimize individual differences in enzyme production and strength, among many factors, use a number of animals for making each extract. Do not wait too long with the digestion experiments. Even in the refrigerator, the extract loses much of its strength in a comparatively short time.

An appropriate buffer must be added in the ratio of 1 buffer:10 extract-substrate mixture to ensure a constant and optimal pH. A constant temperature (30° C. for the frog) should be maintained for the enzyme-substrate mixture undergoing digestion. A thermostatically regulated stove or water bath could be used. Duration of time for digestion should be the same in all experiments. Microtitrations will probably be called for, but these are really not more difficult than macrotitrations, as long as the necessary precautions and procedures are rigorously followed (5).

It is advisable not to analyze too many different enzymes at the same time. The determination of amylase and maltase activity is usually not difficult. Any (micro-) sugar titration will do. (A very good one is that of Hagedorn-Jensen.) Determination of proteases (pepsin and trypsin, for instance) is more difficult and should not be undertaken before experience in titration has been achieved.

A less accurate, but easier method is to introduce a small piece of exposed and developed film into the enzyme-buffer mixture. Its gelatin layer serves as a substrate. Determine the time it takes for the film to become transparent. Compare these digestion times, which gives a clue to the enzyme activities.

To determine relative lengths of the intestinal tract, first measure the total length of the animal minus tail (tadpole) or legs (frog). Then isolate the intestinal tract and measure its length. Express the latter data as a percentage of the former ones.

References · general

(Any laboratory manual on biochemistry will give methods to determine quantitatively sugars, proteins, and their derivatives.)

1. Baldwin, E. 1953. Dynamic aspects of biochemistry. 2d ed. Cambridge University Press, New York.
2. Mitchell, P. 1956. A textbook of general physiology. 5th ed. McGraw-Hill Book Co., New York.
3. Sumner, J. B. and G. F. Somers. 1953. Chemistry and methods of enzymes. 3d ed. Academic Press, New York.

· specific

4. Antonov, V. K., et al. 1970. Investigation of enzyme kinetics by oscillographic polarography. Anal. Biochem. 37(1):160-168.
5. Prosser, C. L. and P. B. van Weel. 1958. Effect of diet on digestive enzymes in midgut gland of African Giant Snail, *Achatina fulica* Fer. Physiol. Zool. 31:171-178.
6. Rowe, W. C., A. K. Huggins, and E. Baldwin. A radio-isoptic assay system for enzymes of the ornithineurea cycle. Anal. Biochem. 35(1): 167-176.
7. Smith, A. C. and P. B. van Weel. 1960. On the protease and amylase production in the midgut gland of young and mature African Snails. Experientia 16:60-61.
8. van Weel, P. B. 1959. The effect of special diets on the digestive processes (enzyme production and resorption) in the African Giant Snail. *Achatina fulica* Bowdich. Z. vergl. Physiol. 42:433-448.

13/EXPERIMENTS IN TISSUE CULTURE

Stanley L. Weinberg
Research and Development Laboratory
John Morrell & Co.
Ottumwa, Iowa 52501

Background

Tissue culture is a method of isolating cells and tissues of multicellular organisms so that particular experiments can be performed that would not be possible in the intact organisms.

With this technique we can grow cells outside the body of the organism, or indefinitely grow tumors which would quickly kill a host.

At present, the knowledge of tissue culture is growing rapidly. Such work is basically simple to do, yet it is easy to make mistakes that will ruin your cultures.

The guiding rules are sterility, carefulness, and persistence.

Suggested Approach

To start tissue culture work, obtain L cell cultures from commercial suppliers or from university tissue culture laboratories. L cells are derived from mice and were first cultured by W. R. Earle at the National Institute of Health. If they have room, the L cells will spread out in a single layer on glass; this is called a monolayer.

Grow cultures in standard 3-oz medicine bottles, or in ordinary test tubes, capped, or stoppered with plastic stoppers. Incubate cultures at 37° C. Slant test tubes at 5 degrees; commercial racks are available for this purpose, but you can support the test tubes on modeling clay or on home-made wooden racks.

Feed cultures with 60% balanced salt solution (BSS—formulas for making up Hank's, Earle's, or Tyrode's are given in the references); 20% embryo extract (purchase—this will be your biggest expense); 20% blood serum or ascitic fluid (get serum free by centrifuging "expired" blood obtained from a blood bank; ascitic fluid can be obtained free from a hospital pathologist, and should also be "spun down").

Add phenol red to the medium, as indicated in the references, to measure pH changes. When making up the medium, add sufficient $NaHCO_3$ to make the medium slightly alkaline—about pH 7.2. Yellowing of the medium means that the cells are metabolizing and producing CO_2, and it is time to change the medium. Use an antibiotic mixture containing penicillin, streptomycin, and mycostatin to keep down contamination.

Cultures can be transferred to new vessels, either for maintenance when the original culture has become overgrown, or to set up new experiments, by "trypsinizing." Pour out the old medium, replace with 0.1% trypsin enzyme in balanced salt solution, incubate for 10 to 20 minutes at 37°C, transfer to a centrifuge tube and spin for 5 minutes at 500 rpm, then remove the supernate; resuspend the "button" of cells in new medium by "trituration" (mixing well with a pipette), then transfer to sterile glassware.

Suggested Problems

If you can get your cells to maintain themselves, this alone is an achievement for a student. When you pass successfully beyond this stage, try one of the following problems:

1. Start primary cultures from chick or mouse embryos. Start primary cultures from adult animals.

2. Try culturing insect or other invertebrate tissues—very little has been done in this area.

3. How do hyaluronidase and versene compare with trypsin (a) as cell-dissociating agents? (b) in their effects on cultures?

4. What effect does light of various wavelengths have on tissue cultures?

5. What effect does heavy centrifugation have on tissue culture? (When transferring cultures, try sedimenting the trypsinized suspension rather than centrifuging it; study the effects of heavy and light centrifugation.)

Possible Pitfalls

Before you begin, buy or borrow one of the manuals listed in the References and read it carefully. Once you start work, maintaining sterility will be your biggest problem. You cannot do tissue culture in an open room with its abundant air currents. If your school does not have a bacterial transfer room, make a hood for tissue culture work out of an old Wardian case, a large fish tank turned on its side, or an old packing case. Routinely check all media and cultures for microbial contamination by using nutrient broth or agar, thioglycollate, and Sabouraud's medium. Use an ultraviolet lamp to sterilize your working surface. Although UV radiation is not very penetrating, the cornea of the eye is easily damaged by it, so that protective goggles (not ordinary eye glasses) should be worn when using the lamp. Another safe procedure is to leave the lamp on overnight, turning it off immediately upon arriving at school in the morning.

Do not be surprised if your first experiments are failures. Do not give up at this stage. Clean up, try to figure out what went wrong, and try again. Other students have successfully mastered tissue culture technique, and you can master it also. But since there are likely to be many false starts and frustrations in tissue culture work, it would be unwise to try to complete a project in this field against a deadline, as for a science fair.

References

1. Cameron, G. 1950. Tissue culture technique. Academic Press, New York.
2. Merchant, D. J. and R. H. Kahn. 1965. The living animal cell/cell and organ culture. Ward's Natural Science Establishment, Inc. Rochester, N. Y.
3. Merchant, D. J., R. H. Kahn, and W. A. Murphy. 1964. Handbook of cell and organ culture. Burgess Publishing Co., Minneapolis, Minn.
4. Parker, R. C. 1961. Methods of tissue culture, 3rd ed. Harper & Row, New York.
5. Paul, J. R. 1970. Cell and tissue culture. 4th ed. The Williams & Wilkins Co., Baltimore.
6. Pollak, O. J. 1970. Tissue cultures. The Williams & Wilkins Co., Baltimore.
7. White, P. R. 1963. The cultivation of plant and animal cells. 2nd ed. The Ronald Press, New York.

Sources of Supply

1. Bellco Glass, Inc., Vineland, N. J. 08630
2. Difco Laboratories, Box 1058A, Detroit, Mich. 48232.
3. Hyland Division, Travenol Laboratories Inc., P. O. Box 2214, 3300 Hyland Ave., Costa Mesa, Calif. 92626.
4. Microbiological Associates, Inc., 4813 Bethesda Ave., Bethesda, Md. 20014.
5. Schwarz BioResearch, Inc., Mountain View Ave., Orangeburg, N. Y. 10962.
6. Ward's Natural Science Establishment, Inc., P. O. Box 1712, Rochester, N. Y. 14603.

14 / FACTORS AFFECTING THE FORMATION AND MOVEMENT OF CARBOHYDRATES INTO THE POTATO TUBER

W. M. Iritani
Department of Horticulture
Washington State University
Pullman, Washington 99163

Background

The potato tuber is a storage organ for carbohydrates manufactured by the leaves and not utilized for vegetative growth. Yield and high dry matter content of potatoes is dependent upon the formation of an adequate photosynthetic leaf area, after which vegetative growth decreases or ceases and the carbohydrates are translocated into the tubers. After tubers are initiated they compete with the vegetative portion for carbohydrates manufactured by the leaves. That is, if environmental conditions are favorable for vegetative growth, then very little tuber development occurs. If vegetative growth is limited by such factors as shortening day length and cooler temperatures, rapid tuber development takes place.

Although some knowledge is available concerning factors that control vegetative growth and tuber development, it is insufficient to be of practical value. More basic knowledge of these factors would be a significant contribution to the science of food production. The production of high quality potatoes is dependent upon the ability of a plant to manufacture and translocate carbohydrates into the tuber.

Suggested Approach

It is generally known that levels of various nutrients have an effect on the growth processes. For example, high nitrogen levels tend to produce vegetative growth at the expense of carbohydrate movement into the tuber. Photoperiod and temperature during the various periods of the growing season also have an effect. Although these last two factors can be controlled to some extent by altering cultural practices, it is not sufficient to be of significant value.

The most promising method of regulating the growth mechanisms seems to lie in the use of growth hormones. Both growth-promoting as well as growth-depressing hormones could be used in designing experiments. For example, the growth-promoting hormones could be used to increase photo-synthetic leaf area early in the season, growth-depressing hormones could be used to depress vegetative growth later on and promote the movement of carbohydrates into the tubers. Gibberellic acid has been reported to hasten the emergence of plants in the spring, thus promoting vegetative development early in the season. On the other hand, excessive vine growth later on which can result from high temperatures could possibly be controlled by the use of growth-depressing regulants such as 2-chloroethyl trimethyl ammonium chloride (CCC).

Every year new growth regulators are being formulated which offer exciting new areas of research in plant growth regulation. Many experiments can be designed to test the effect of these hormones. Investigations on optimum concentrations to use as well as time of application will need to be conducted.

References · general

1. Audus, L. J. 1960. Plant growth substances. 2d ed. John Wiley & Sons, New York.
2. Bonner, J. F. and A. W. Galston. 1952. Principles of plant physiology. W. H. Freeman & Co., San Francisco.
3. Ferry, J. F. and H. S. Ward. 1959. Fundamentals of plant physiology. The Macmillan Co.., New York.
4. Mitchell, J. W. and P. C. Marth. 1947. Growth regulators for garden, field, and orchard. University of Chicago Press, Chicago.
5. Skoog, F. 1951. Plant growth substances. University of Wisconsin Press, Madison.
6. Went, F. W. and K. V. Thimann. 1937. Phytohormones. The Macmillan Co., New York.

· specific

7. Burton, W. G. 1948. The potato. Chapman & Hall, Ltd., London.
8. _____ and N. G. Wiggington. 1970. The effect of a film of water upon the oxygen status of a potato tuber. Potato Res. 13(3):180-186.
9. Burton, W. G. and A. R. Wilson. 1970. The apparent effect of the latitude of the place of cultivation upon the sugar content of potatoes grown in Great Britain. Potato Res. 13(4):369-383.

10. Ohms, Richard E. 1962. Producing the Idaho potato. Idaho Agr. Extension Serv. Bull. 367, February.

11. Terman, G. L. 1950. The effect of rate and source of potash on yield and starch content of potatoes. Mine Agr. Expt. Sta. Bull. 481.

12. _____, et al. 1951. Rate, placement, and source of nitrogen for potatoes in Maine. Maine Agr. Expt. Sta. Bull. 490.

13. Werner, H. O. 1947. Commercial potato production in Nebraska. Agr. Expt. Sta. Univ. of Nebraska Bull. 384.

15 / EFFECTS OF METALS ON TRANSPORT IN *ELODEA CANADENSIS*

Background

Benjamin Lowenhaupt
Department of Biology
Edinboro State College
Edinboro, Pennsylvania 16412

One of the most important unsolved problems of biology is the mechanism of active transport, that is, the metabolic transfer of material against the electrochemical potential gradient. Examples of active transport are the secretion of acid in the stomach, the uptake of salt and sugar in the kidney, and the accumulation of ions by plant roots.

The mechanics of transport and the effects of metals on transport can be studied advantageously in the leaves of certain aquatic plants. These leaves transport calcium in the light; they accumulate calcium from the medium into their morphological under side and excrete it back into the medium from their morphological upper side. As a result, the pH of the medium becomes inhomogeneous; the medium adjacent to the upper leaf surface becomes alkaline. This fact makes possible a simple and unambiguous test for transport.

Metal ions that inhibit transport include copper, zinc, lead, uranium, magnesium, sodium, and potassium. But calcium in the medium seems to counteract the inhibiting effect because some of the listed ions do not inhibit transport when there is abundant calcium in the medium. This suggests that calcium and these ions compete for a molecular site in the leaves, that calcium attachment to the site is necessary for transport and when another ion displaces the calcium, transport is inhibited. Some of the listed metals inhibit transport even when there is considerable calcium in the medium. Accordingly it appears that these ions compete for the molecular site very strongly, that they can displace calcium even when they are at low concentrations and the calcium is relatively abundant.

Statement of the Problem

The suggested explanation of transport inhibition can be tested by studying the inhibition quantitatively. If our postulate is correct, there should be a quantitative relationship between the concentration of calcium in the ambient medium and the minimum concentration of the inhibiting ion to prevent transport. To test for this relationship, you may find out how much of the inhibiting ion is needed to block transport at different calcium concentrations. Increasing the calcium in the medium or removing some of the inhibiting ion should restore transport. From your data try to find a formula relating calcium concentration and the inhibiting potency of the metal ion. This formula may contain the ratio:

$$(calcium)^m/(inhibitor)^n$$

where the parentheses indicate concentrations of calcium and inhibitor, respectively, and "m" and "n" are exponents that you must determine from the data.

From this formula it may be possible to infer something about the chemical kinetics of the inhibition. But to make this analysis you will need help from a chemist or biochemist.

Whether it is possible to express inhibition in terms of the concentration of the ion in the surrounding medium must be ascertained, because if this ion has a strong chemical affinity for the biological material its concentration in the medium will decline during the experiment. Perhaps to use a very large volume of solution and only a little biological material will prevent a concentration change, or perhaps you can use flowing solution. If nothing else is possible, express the inhibition as a function of the total amount of the poisonous ion, rather than its concentration. But this will give less useful results. (To ascertain whether the ion is depleted from the medium, a number of tests can be devised. Biological tests may be satisfactory, but the radio-tracer method is suggested if it is available.)

Suggested Approach

The kind of aquatic plant to use in these experiments may depend upon what is available. The leaves of many freshwater plants transport ions and probably any of these would be satisfactory. In fact, to list species with this property would be an interesting and useful project, which has not been done in the United States (but see Ruttner, 1947). I suggest *Elodea canadensis* for the experiments because it is easy to grow and it often can be obtained from commercial growers. Florists call it Anacharis. Plants from commercial growers are excellent if they are used when fresh. Plants sold by aquarium supply shops or shipped long distances are often useless except to start your own planting.

Elodea grows well in 55-gallon steel barrels, painted inside to prevent rust. Use about six inches of rich soil at the bottom of the barrel. Add water and allow to settle until the water is not very cloudy. Then thrust the base of *Elodea* cuttings into the soil and hold there with a weight. These cuttings will root and grow quickly. (It may help to put duckweed in the barrel, since it seems to discourage algae, but do not let it shade out the *Elodea*.) When the *Elodea* is well established add some powdered $CaCO_3$ to the water, a goldfish, and some victory snails. You must restart the *Elodea* from time to time.

To test for transport, stir some powdered $CaCO_3$ in the water, allow to settle, and decant the supernatant into a flask. Add phenolphthalein dissolved in ethanol (alcohol) to this supernatant in sufficient amount that an aliquot becomes deep red when NaOH is added. Discard the aliquot with NaOH. Use a concentrated solution of phenolphthalein so that only a little alcohol needs to be added. The $CaCO_3$-phenolphthalein solution should be colorless. If it is pink, bubble your breath through it until the color fades. A sprig of *Elodea* is placed in a large test tube, rinsed with water, and covered with the solution. Stand in the light for a few minutes. A red cloud at the upper surface of each leaf indicates transport.

Possible Pitfalls

1. All water should be glass distilled and stored in glass. Both ordinary distilled water and tap water are often poisonous. But tap water can be used to grow the plants. (I suppose this is because the soil detoxifies it.)

2. Use only healthy and clean leaves. Epiphytes and encrusted dirt cannot be removed; they can only be avoided by using young plants vigorously growing in clean water. Above all, do not waste your time with diseased plants.

3. To consider toxicity in terms of concentration has theoretical objections that you must bear in mind. The *activity* rather than the *concentration* may be important, and at moderate concentrations these may be sig-

nificantly different. However, you will probably use only very dilute solutions—such low concentrations that the activities and concentrations may be considered the same.

4. Remember that the acidity of the medium during a test may be critical. To investigate the effects of acidity would be a difficult project. I suggest that you buffer the medium with tris (hydroxymethyl) amino methane, adding HCl to it until a final pH of 8 is reached.

References · general

1. Bonner, J. F. and A. W. Galston. 1952. Principles of plant physiology. W. H. Freeman & Co., San Francisco.
2. Crafts, A. S. 1961. Translocation in plants. Holt, Rinehart & Winston, New York.
3. Davson, H. 1959. A textbook of general physiology. 2d ed. Little, Brown & Co., Boston. (To conduct and interpret your research intelligently you must think about it in the broadest possible physiological terms. This book will start you.)
4. Galston, A. W. 1961. The life of the green plant. Prentice-Hall, Inc., Englewood Cliffs, N. J.
5. Hoagland, D. R. 1948. Lectures on the inorganic nutrition of plants. The Ronald Press Co., New York. (This is the best general reference on transport in plants that I know.)

· specific

6. Davson, H. and M. B. Segal. 1970. The effects of some inhibitors and accelerators of sodium transport on the turnover of ^{22}Na in the cerebrospinal fluid and the brain. J. of Physiol. 209:131-153.
7. Lowenhaupt, B. 1969. Bioelectrochemical throttle of metabolism. J. of Theor. Biol. 25(2):187-207.
8. _____. 1970. Vital staining by Fe and Mn salts of electron microscopy of plant tissue (*Elodea*). Stain Tech. 45(1):29-34.
9. Robertson, R. N. 1958. The uptake of minerals. In W. Ruhland, Encyclopedia of plant physiology, Vol. 4, Mineral nutrition of plants, pp. 243-279, Springer-Verlag, Berlin.

16 / THE ROLE OF LIGHT AND TEMPERATURE IN THE SEASONAL DISTRIBUTION OF MICROCRUSTACEA

Kenneth B. Armitage
Department of Biology
University of Kansas
Lawrence, Kansas 66044

Background

Ponds and lakes abound with free-swimming Crustacea, of which the Cladocera and Copepoda are the most widely distributed. Some species are perennial; others are seasonal. The seasonal species are typically vernal or autumnal. Also, perennial species may have seasonal peaks of abundance. Little information is available concerning the environmental factors that control the seasonal distribution of these animals. Food is an obvious factor. Indeed, laboratory studies have indicated that population fluctuation is related to changes in food supply. But since most of these crustaceans are filter feeders and probably eat any algae of the correct size, it seems unlikely that food directly controls the seasonal distribution.

Two environmental factors that demonstrate seasonal patterns are light and temperature. In the spring in the northern hemisphere, both the length of day and the temperature of the water increase. In the autumn both day-length and temperature decrease. This combination of temperature and light results in four light-temperature patterns. For example, autumnal species are present during cool temperatures and decreasing day-length; winter species are present during cold temperature and short day-length (decreasing until December 21, but increasing thereafter). Day-length influences the reproductive cycles of animals, and temperature controls the distribution of animals. Thus the interaction of these two environmental factors seems to determine the patterns of seasonal distribution of many species of microcrustaceans.

Several studies in recent years indicate that such an effect by temperature and light may indeed occur. Cultures of Simocephalus vetulus, a cladoceran, were kept in constant light and constant dark. Animals cultured in constant light had a much higher reproductive rate than the animals cultured in constant dark. The reproductive cycles of some marine invertebrates are correlated with increase in day-length or with fall in temperature. In the life cycles of some of these crustaceans there is a period of arrested development or diapause. Diapause in *Diacyclops navus* was caused primarily by day-length but the effect of day-length was modified by temperature. Diapause in *Daphnia* may be modified by the density of animals. Temperature also affects the metabolic rates of all microcrustaceans that have been measured. This effect of temperature might be translated directly into population growth or suppression of population growth. Studies of oxygen consumption of marine copepods indicate that the animals have a high rate of respiration during the time of their maximum abundance.

Suggested Problems

Aquatic biologists have determined the seasonal distribution of many microcrustacea and other zooplankton forms in various regions. The student's problem is to determine the patterns of seasonal distribution as correlated with light and temperature changes.

Suggested Approach

There are two essentials necessary for pursuing studies correlating the role of light and temperature with seasonal distribution. First, the seasonal distribution of the organism(s) to be studied must be known. Secondly, it is necessary to grow the animal being studied in the laboratory. Culturing microcrustacea can be tricky. Brewer's yeast may be used, but unicellular algae seem to be better. Several of the species of *Chlamydomonas* are readily cultured on media in the laboratory and may be used to feed the crustaceans. Once crustacea are successfully cultured, they may be reared under various light conditions.

A simple experiment involves rearing cultures under 12-16 hours of light per day while treating other cultures with about 8 hours of light per day. Each culture should contain about 50 cc of filtered pond water as the culture medium. At least 10 replicate cultures are maintained under each condition of light. The animals in each culture are counted weekly and placed in fresh medium. Since reproduction is of primary concern, the adults and juveniles are enumerated separately. Care must be taken to give each culture the same amount of food. Not all microcrustaceans culture easily and some of the references should be consulted for culturing techniques.

The experimental setup just described can be maintained at any temperature. Room temperature is usually about 22°C. An organism cultured at this temperature might be expected to do well. Experiments may be run at 5- or 10-degree (C) intervals above and below room temperature.

If a temperature effect on population growth is found, oxygen consumption studies may be made and an R/T (rate/temperature) curve constructed. In this type of study it is important to use animals close to the same size for all the determinations of oxygen consumption. Ideally, oxygen consumption should be determined at 3° C intervals, from 2° C to 35° C or higher, depending on the responses of the animal. Of course, it is possible that animals that inhabit cold water may die before high temperatures are reached. Conversely, warm-water animals may die at low temperatures. Many organisms are capable of some degree of adjustment to temperature. Usually some period of time is necessary before adjustment occurs. Such an adjustment to temperature with time is known as acclimation. The experimental animals should be acclimated to each temperature before oxygen consumption is determined. Acclimation is easily accomplished by slowly raising or lowering the temperature of the medium in which the animals are kept. The temperature change should not be more rapid than 1°C per day; the acclimated situation more closely approximates natural conditions of ponds and lakes where temperature of the water changes slowly.

Possible Pitfalls

Probably the chief problem in conducting a series of experiments of the sort described is the control of temperature. Some sort of water bath is necessary for maintaining a constant temperature during the period when oxygen determinations are being made. Water baths may be improvised if a large container and hot and cold water are available. The more critical problem will likely have to do with long periods of time needed for acclimating animals, or for population-growth studies. A small room with a large air conditioner may serve to maintain temperatures as low as the 10°-15° C range. If such temperature control is not available, oxygen consumption may be measured under semiacclimated or acute conditions. Determination of oxygen consumption under semiacclimated or acute conditions will indicate, in part, the ability of an animal to flourish at a given temperature.

Oxygen consumption can be determined with a high degree of accuracy by the Winkler method. However, it is necessary to practice making oxygen determinations on replicate samples until a high level of precision is attained. Light is easily controlled. A clock mechanism, such as the type used in chicken coops, can be attached to any ordinary light switch and set to turn the lights on and off at prescribed times.

References · general

1. Edmondson, W. T., ed. 1959. Freshwater biology. 2d ed. John Wiley & Sons, New York.
2. Hutchinson, G. E. 1967. A treatise on limnology. II. Introduction to lake biology and the limnoplankton. John Wiley & Sons, Inc., N.Y.

· specific

3. Bullock, T. H. 1955. Compensation for temperature in the metabolism and activity of poikilotherms. Biol. Rev. 30:311-342.
4. Conover, R. J. 1959. Regional and seasonal variation in the respira- of marine copepods. Limnol. Oceanog. 4:259-268.
5. Giese, A. C. 1959. Reproductive cycles of some West Coast invertebrates. In R. B. Withrow, ed., Photoperiodism, AAAS Publ. No. 55: 625-638.
6. Heinle, D. R. 1969. Temperature and zooplankton. Chesapeake Science 10:186-209.
7. Parker, R. A. 1959. *Simocephalus* reproduction and illumination. Ecology 40:514.
8. Siefken, M. and K. B. Armitage. 1968. Seasonal variation in metabolism and organic nutrients in three Diaptomus (Crustacea: Copepoda). Comp. Biochem. Physiol. 24:591-609.
9. Standard methods for the examination of water, sewage and industrial wastes. 1970. American Public Health Association, New York.
10. Stross, R. G. 1969. Photoperiod control of diapause in *Daphnia*. III. Two-stimulus control of long-day, short-day induction. Biol. Bull. 137:359-374.
11. Waterman, T. H., ed. 1960. The physiology of Crustacea. Vol.1. Metabolism and Growth. Academic Press, New York.
12. Watson, N. H. F. and B. N. Smallman. 1971. The role of photoperiod and temperature in the induction and termination of an arrested development in two species of freshwater cyclopid copepods. Can. J. Zool. 49:855-862.

17 / SUSPENDED SEDIMENT AS AN ECOLOGICAL FACTOR IN THE BEHAVIOR OF THE OYSTER BORING GASTROPOD *UROSALPINX CINEREA*

Melbourne R. Carriker
Director, Systematics-Ecology Program
Marine Biological Laboratory
Woods Hole, Massachusetts 02543

Background

The marine oyster borer *Urosalpinx cinerea* is a dominant predator in many intertidal and subtidal hard-bottom and oyster communities along the east coast and some regions of the west coast of the United States, and in Great Britain (1, 5). Large populations living in or close to the bottom, especially inside estuaries, find themselves in a gradient between sediment and water exposed to periodic pelting by a rain of particles raised by ebbing and flooding tides and by storms (2, 7). Grains from this aquatic dust storm settle on the external integument of the snail and others are transported into the mantle cavity by water currents maintained for respiratory purposes by beating cilia on gills and mantle surfaces. The epidermis on the external soft parts of the mollusk, including the mantle cavity, contains numerous mucous and ciliated cells that function continuously as a sanitation system to remove these sediments. Mucous, secreted in a thin sticky sheet to which the particles adhere, acts in conveyor-belt fashion as a transport medium that is then moved off the snail by the cilia (4).

By means of chemosensory receptors, probably within the osphradium in the mantle cavity and on the anterior surface of the foot, this gastropod detects a scentlike substance (as yet unidentified) from food organisms (a number of species of barnacles and epifaunal bivalves) (15). It crawls over the bottom in the direction of the source of the attractant toward the prey. Once a prey—an oyster, for example—is contacted, the snail creeps onto it, bores a hole through its calcareous shell by chemomechanical means, and rasps out the soft flesh within (8, 10, 13, 14). The snail's ciliary-mucous cleansing mechanism undoubtedly facilitates detection of prey by freeing its chemosensory organs of foreign particles.

Female oyster borers oviposit fertilized ova in vase-shaped egg capsules molded by a pouchlike gland in the anterior midventral surface of the foot. Ova pass into this ventral pedal gland from the genital pore of the female by way of a ciliated groove over the right side of the foot of the snail (4). Unimpeded passage of clean ova to the ventral pedal gland is probably insured by synchronous activity of mucous and ciliated epidermal cells.

Statement of the Problem

How heavy loads of sediment roiled into the water by swift tidal currents and storms affect respiration, creeping, detection of prey, boring of shell, feeding, oviposition, and development of young snails in the egg capsules has not been tested experimentally. Knowledge of this kind would be a valuable addition to the fundamental autecology and behavior of gastropods, and might suggest means of controlling this serious pest of oysters and clams. It is conceivable, for example, that natural clay minerals in coastal waters may interfere by adsorption with the capacity of these snails to detect chemical signals from prey. Incidently in the same way suspended sediments may ameliorate the deleterious effects on coastal organisms of the varied poisons which increasingly are being dumped into coastal waters by man. National interest in the effects of suspended and deposited sediments on estuarine and coastal organisms has been heightened by the need to establish national water quality criteria and standards and by coastal engineering projects which result in temporary or enduring changes in suspended loads and deposition of sediments (6).

It is suggested that the effects of a range of concentrations of (a) fine natural sediments (silts and clays) and of (b) montmorillonite (clay), particle sizes ranging under 20 , be tested on the activities of populations of 50 oyster borers at a time (25 females and 25 males, in the shell height range of 20 to 35 mm) in 20 liters of seawater (salinity 25 to 32 parts per thousand) preferably similar to the average of the salinity in which the snails lived natively), in closed recirculating aquaria at 20° C. Snails in similar aquaria free of sediments should be used as controls. Any one, or all, of the following activities of snails in each experimental tank, contrasted with those in the control tanks, may be used as quantitative indices of the effects of the suspensions after given period of exposure.

(a) number of snails crawling out of the water on the sides of aquaria per unit time;

(b) rapidity (timed with a stop watch) with which snails right themselves when placed on their backs;

(c) time for, and percentage of, snails placed at one end of the aquarium to crawl a given distance onto 5 freshly collected active oysters (approximately 2.5 cm long) at the opposite end of the tank;

(d) time for snails to bore and start feeding on oysters, indicated by partial gaping of valves of oysters;

(e) choice of oviposition site in aquarium by females, rate of oviposition of egg capsules, and time for emergence of young snails from the capsules;

(f) mortality rate, particularly after prolonged exposure to dense concentrations of suspended sediment.

Experiments should be carried out for a minimum of one week. During the first week, observations on (a) could be made daily or twice daily, and snails out of water on the sides of the tank returned to the bottom of the aquarium. At the end of the week, experiments (b) and (c) could be carried out and observations on (e) (first two aspects) and (f) recorded. Oysters in (c) should be left in aquaria a minimum of 6 hours. Since oysters when pumping quickly filter suspended sediment out of the water (13), it will be necessary by visual means (or a turbidimeter) to determine the rate of loss and add appropriate quantities to maintain the initial concentration of sediment in the water for the duration of the observation. In experiment (d) which is a continuation of (c), maintain the initial concentration of sediment in the experimental aquarium, and observe when gaping of oysters takes place. In the third aspect of (e), likewise maintain the initial concentration of sediment in the experimental tank and observe when young snails emerge (1).

Suggested Approach

Research of this kind is best conducted within relatively short distances of the seacoast. Unpolluted seawater, checked with a hydrometer to determine whether the salinity is sufficiently high, should be taken at high tide near the ocean where high salinities prevail. Seawater may be hauled and stored indefinitely in 5-gallon polyethylene or glass screw-cap bottles. The pH of seawater stored in this way, however, soon drops (check the pH with a pH meter or indicator paper), and should be alkalized with sodium bicarbonate to a pH of 8.0 before use. Adding broken mollusk shell to the bottles will help prevent excessive acidification of the seawater. Oysters and oyster borers may be gathered along the coast during low tide on pilings, rock jetties, and oyster clusters. Oysters may be transported out of water, but oyster borers should be carried in just enough water to cover them.

In the laboratory these animals may be maintained for long periods of time in a large covered aquarium in recirculated seawater changed monthly or bimonthly. As with the bottled seawater, increase the pH as required. A few oysters at a time should be provided the snails for food. The stock supply of oysters may be held on a platform suspended within the oyster borer tank, but out of reach of the snails. Oysters will clean the seawater by ciliary filtration and thus obviate the need of mechanical filtration. Recirculation is achieved by passing seawater from the lower side of one end of the aquarium at table height through rubber and glass tubing about 12 mm in diameter to a large glass "Y" tube near the floor where compressed air is injected through a tapered glass nozzle into one arm of the "Y," and thence up into the aquarium again at the opposite end. Compressed air (whether from a central supply or a small pump) should be passed through cotton and bubbled through tap water before use to remove oil and other impurities. The outlet end of the recirculating line in the aquarium is covered with a section of rubber tubing stoppered at the outer end and perforated with a series of narrow slits to permit flow of water, but to exclude oyster borers. Aquaria are covered by a snug lid to reduce evaporation (3).

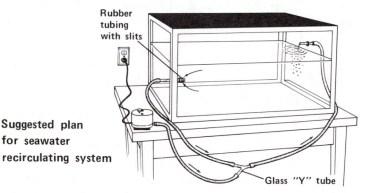

Suggested plan for seawater recirculating system

Rubber tubing with slits

Glass "Y" tube

Possible Pitfalls

All aquaria, tubing, and other containers or surfaces used to hold live animals should be scrupulously clean and leached of possible toxic chemicals by soaking in running tap or seawater for several days before use. Contaminated containers are cleaned by scrubbing gently with fine sand, washing alternately with 10% HCl and 70% ethyl alcohol, and finally thoroughly rinsing in running tap water. Where it is possible live animals should be maintained in fresh running seawater.

For experimental tanks, covered Plexiglas® aquaria are most satisfactory. A size of 30 x 30 x 60 cm is convenient, and recirculation of water may be patterned after that described for the stock tank. The number of tanks employed at a time will depend on the number of concurrent experiments and observations; a minimum of three tanks is envisioned: a control aquarium, and two aquaria for duplicate tests of each size and kind of sediment.

If the salinity of the seawater to be used in the tanks is higher than the average of that in which the snails were collected (as determined by hydrometer or silver nitrate titration), it should be diluted with tap water to provide the appropriate salinity. Concentrations of one gram per liter of sediment should be employed first. The density may then be increased one gram at a time in subsequent tests (11, 16, 17). Montmorillonite may be obtained commercially. Fine natural sediment may be collected in quiet bays along the coast. This may be dried in an oven and then sieved to obtain the desired sizes of silt and clay particles. Check the range of size of the particles with a micrometer scale in a compound microscope. Whereas montmorillonite will be a relatively pure clay mineral, the natural sediment will consist of a combination of various silts and clays and probably some organic matter (7). Determine the degree of mixing of seawater necessary to maintain the sediments in suspension in the experimental tanks. For low concentrations of clay, flow from the recirculation of seawater may suffice. For heavier grains, install one or more laboratory stirrers; extend the stem of the stirrer through an opening in the Plexiglas® cover of each tank on supports insulated from the aquarium so that vibration from the stirrers and motors may be reduced and transmitted minimally to the animals. The effect of excessive vibration on these animals is uncertain, and may confuse the results of the experiments.

To determine the sex of the oyster borer, allow the snail to crawl onto one of your fingers immersed in seawater, and then gently and slowly roll the shell to the animal's left, exposing the anterior right portion of the mantle cavity. In males a long slender tapering partly coiled penis will be seen just behind the right tentacle; females generally lack this, or may possess only a vestigial stump. If you encounter intermediate sized penises, do not use the snails as their sex is indeterminate.

Dead oyster borers are withdrawn deeply into the shell and elicit no response when the soft parts beyond the operculum are pricked with a dissecting needle. In snails dead a day or so the operculum readily peels off the withdrawn soft parts.

Oyster borers may climb the sides of aquaria even in the presence of adequate food and clean aerated seawater. Why this is so is not known, but may be related to existence in the vicinity of the intertidal zone. For more exact experimentation, it would be desirable to carry out these studies in laboratory tide models (18). Under less tolerable conditions their climbing behavior is accelerated. As the concentration of suspended sediments is increased, it may be necessary to return snails to the bottom of their tanks several times a day.

Snails in both experimental and control tanks may survive starvation diets for six months or more. It is recommended, however, that normally feeding oyster borers be used in these experiments, that is, those fed every week or two. They may be fed opened oysters to reduce the time required for them to bore.

References · general

1. Carriker, M. R. 1955. Critical review of biology and control of oyster drills *Urosalpinx* and *Eupleura*. U. S. Fish Wildl. Serv., Spec. Sci. Rept. Fish. 148:1-150.
2. _____. 1967. Ecology of estuarine benthic invertebrates: a perspective. In G. H. Lauff, ed., Estuaries, Amer. Assoc. Adv. Sci., Publ. No. 83, 442-487.
3. Clark, J. R. and R. L. Clark. 1964. Sea-water systems for experimental aquariums. A collection of papers. Fish & Wildl. Serv., Res. Rept. 63.
4. Fretter, V. and A. Graham. 1962. British prosobranch molluscs: their functional anatomy and ecology. Royal Society, London.
5. Hancock, D. A. 1959. The biology and control of the American whelk tingle *Urosalpinx cinerea* (Say) on English oyster beds. Great Britain Min. Agr. Fish. Food, Fishery Invest. Ser. II, 22(10):66.

6. Sherk, J. A., Jr. and L. E. Cronin. 1970. The effects of suspended and deposited sediments on estuarine organisms. An annotated bibliography of selected references. Nat. Res. Inst., Univ. Md., Ref. No. 70:19.
7. Shepard, F. P. 1963. Submarine geology, 2d ed. Harper & Row, Publishers, N. Y.

· specific
8. Carriker, M. R. 1943. On the structure and function of the proboscis in the common oyster drill, *Urosalpinx cinerea* (Say). J. Morph. 73: 441-506.
9. _____. 1969. Excavation of boreholes by the gastropod, *Urosalpinx:* an analysis by light and scanning electron microscopy. Amer. Zool. 9:917-933.
10. _____ and D. VanZandt. 1971. Shell penetrating behavior of a predatory boring muricid gastropod. In H. E. Winn and B. L. Olla, eds., Behavior of marine animals—recent advances. Plenum Press (in press).
11. Davis, H. C. 1960. Effects of turbidity-producing materials in sea water on eggs and larvae of the clam (*Venus [Mercenaria]* mercenaria). Biol. Bull. 118:48-54.
12. _____ and H. Hidu. 1969. Effects of turbidity-producing substances in sea water on eggs and larvae of three genera of bivalve mollusks. Veliger 11:316-323.
13. Galtsoff, P. S. 1964. The American oyster *Crassostrea virginica* Gmelin. Fish. Bull., Fish Wildl. Serv. Vol. 64.
14. Hanks, J. E. 1957. The rate of feeding of the common oyster drill, *Urosalpinx cinerea* (Say), at controlled water temperatures. Biol. Bull. 112:330-335.
15. Kohn, A. J. 1961. Chemoreception in gastropod molluscs. Amer. Zool. 1:291-308.
16. Loosanoff, V. L. 1961. Effects of turbidity on some larval and adult bivalves. Proc. Gulf and Caribbean Fish. Inst. 14:80-95.
17. _____ and F. D. Tommers. 1948. Effect of suspended silt and other substances on rate of feeding of oysters. Science 107(2768):69-70.
18. Thompson, T. E. 1968. Biology notes. Experiments with molluscs on the shore and in a laboratory tide-model. School Science Review (Great Britain), No. 170:1-6.

18/POPULATIONS OF MILKWEED BEETLES

Richard A. Edgren
Warner-Lambert Research Institute
Morris Plains, New Jersey 07110

Background

In the Chicago region during late June, the conspicuous red and black milkweed beetle (*Tetraopes tetraophthalmus*) begins to appear on plants of the common milkweed (*Asclepias syriaca*). In this area the population rises to peak density toward the end of July and then begins to decline, the last of the beetles dying off about the second week in September. A second species (*T. femoratis*) occurs in the area, but it does not appear until late July and only represents a small proportion of the total beetle population. The relatively large size, conspicuous coloration, sedentary habits, and negative geotaxis of the milkweed beetles make them very suitable animals for the study of changes in population density in the field. Also, their wide distribution in North America makes them readily available for study in many areas. Movements of the beetles from one locality to another can also be studied.

Suggested Problems

How, through the course of a single season, does population density vary in milkweed beetles? Does density change from year to year? Is there more than one species present in any single locality? If so, do they differ in population density through the year? Does one species of beetle always select the same species of milkweed for food? (There are thirteen species of *Tetraopes* in the United States; the species may be identified by the use of Reference 5.) Usually it is impossible to count the animals in a population, so estimates of density of population must be made by an indirect technique. The beetles are marked so that they can be identified on recapture. A spot of model-airplane glue (or nail polish) on the thorax or hard outer wings will serve to identify a beetle when it is recaptured. When population estimates are desired the beetles captured on one day may be marked with one color, and a different color may be used for each subsequent day when collection is carried out. When movement is to be studied, it is necessary to identify individual beetles, and an individual marking system is necessary. The following is a five-color code which has been used successfully to identify five hundred individuals. In this code, spots of color are placed on different parts of the beetle. The table indicates a number value for each color and for each spot. By adding all the numbers representing the colors and spots, one can find a number identifying each beetle.

	Color 1	Color 2	Color 3	Color 4	Color 5
Spot on Thorax	100	200	300	400	500
Spot on Wing Cover					
Left Wing Anterior	10	30	50	70	90
Left Wing Posterior	20	40	60	80	—
Right Wing Anterior	1	3	5	7	9
Right Wing Posterior	2	4	6	8	—

Thus, a beetle marked according to the following scheme is Beetle 158.

Color 1 on thorax	=	100
Color 3 on Anterior of left wing	=	50
Color 4 on Posterior of right wing	=	8
		158

After a series has been marked, the number of beetles in the population may be estimated from data collected on a second field trip by using what is called the Lincoln index:

$$\frac{\text{Total Marked Beetles}}{\text{Total Beetles in Population}} = \frac{\text{Total Marked Beetles in Sample}}{\text{Total Beetles in Sample}}$$

For example: On July 8, 1972, 502 milkweed beetles were captured; of these 162 had been marked on July 7, at which time a total of 411 beetles had been captured, marked, and released back into the colony.

Thus:

$$\frac{411}{X} = \frac{162}{502}$$

Solving for X, the population was calculated to be 1273 or, in round numbers, about 1300 beetles.

If movement of animals within a population is to be examined, the study area may be divided into sectors, and shifts in position of beetles from sector to sector may be recorded. In a Chicago area population that was studied intensively (two trips daily for two weeks) it was found that certain individuals moved habitually (an average of about once a day), whereas others remained in isolated patches of plants, often on the same plant, throughout the study period.

Pitfalls

Both the animals and the method of estimate must be handled with care. The beetles will fight if confined in close quarters. A collecting box or jar that contains too many beetles is usually also filled with amputated legs and antennae. A satisfactory method follows. Some 20 to 30 beetles are captured and placed in a box (cigarette boxes are excellent). They are marked and released at the same time. This procedure is repeated until the entire study area has been covered. This also largely obviates a difficulty in the method of estimate. The accuracy of the Lincoln index depends upon the random dispersal of marked animals into the population after release. If all marked animals are released in one corner of a field of milkweed, one would expect more to be concentrated there on a return trip, thus perhaps affecting the accuracy of the count.

Some beetles die and others fly away, so it is impossible to keep adding the number of beetles marked on subsequent trips to those marked earlier to build up a high number of "Total Marked Beetles." Our studies have allowed us to estimate the half-life of marked beetles at 9 days and this was used as a correction for the "Total." In many of our own studies we have marked on Saturday and used beetles recaptured on Sunday to estimate the population. Death and emigration can probably be ignored in these circumstances; however, we do not know whether this 24-hour period is long enough for the beetles to become randomly dispersed into the population. At any rate, a satisfactory index of population size can be obtained using these methods.

It is suggested that the population of a given area be estimated at least once every two weeks through one or more summers. In selecting an area to study, it is important to attempt to use an isolated area of milkweed in order to reduce the probability of emigration or immigration.

References · general

1. Andrewartha, H. G. and L. C. Birch. 1954. The distribution and abundance of animals. University of Chicago Press, Chicago.
2. Lack, D. 1954. The natural regulation of animal numbers. Oxford University Press, New York.
3. Pearl, R. 1930. The biology of population growth. The Macmillan Co., New York.
4. Warren, K. B., ed. 1957. Population studies: animal ecology and demography. Cold Spring Harbor Symposia on Quantitative Biology, Vol. XXII, Cold Spring Harbor, N. Y.

· specific

5. Chemsak, J. A. 1963. Taxonomy and bionomics of the genus *Tetraopes* (Cerambycidae: Coleoptera). Univ. California Publ. in Entomol. 30:1-90. University of California Press, Berkeley.
6. Nicholson, A. J. 1933. The balance of animal populations. J. Animal Ecol. 2(3):132-178.
7. Solomon, M. E. 1949. The natural control of animal populations. J. Animal Ecol. 18(1):1-35.

19 / SELECTIVE DEFAUNATION OF THE TERMITE HIND-GUT AND CULTIVATION OF TERMITE PROTOZOA

Rolf Mannesmann
Texas Forest Products Laboratory
Lufkin, Texas 75901

Background

There is a very close relationship between most termite species and the symbionts living in their hind-gut (1). The ability to metabolize cellulose into physiologically useful compounds is based on the cellulase production of the hind-gut microorganisms. Host cellulase activity has been found only in a few cases. Thus, the loss of symbionts generally causes the death of the host (4). It is possible to defaunate the termite hind-gut, for instance, by raising the environmental temperature or pressure, or by placing termites in a high oxygen atmosphere for a short period (5).

Although the knowledge about this symbiotic system is quite extensive, little is known about cellulose decomposition by the different groups of protozoa, bacteria, and spirochaetes found in a termite gut. The question still remains whether all of these organisms take part in cellulose decomposition, and also whether bacteria living within some termite protozoa produce cellulase. Furthermore, the rate of cellulase production by different protozoa species requires further evaluation.

Two different methods may be used to study some of these problems: *Method A.* Evaluation of the termite feeding rate and the cellulose content of the termite feces after partial defaunation of the hind-gut.

Selected organisms, such as either the whole group of flagellates (bacteria, spirochaetes) or single species of these groups which remain in the hind-gut. To eliminate hind-gut organisms selectively, different antibiotics and sulfanilamids may be used. Although this has been tried several times, results, thus far, show little success (8). A second method would include: *Method B.* Cultivation of hind-gut microorganisms *in vitro*.

The determination of the activity of isolated microorganism groups (or species) in cellulose decomposition is necessary. Xylophagous flagellates of the termite gut have been cultivated in only a few cases (10). Different mediums such as buffered nutrient solutions or agar with organic and/or inorganic compounds may be used to get *in vitro* cultures and subcultures.

Suggested Approach for Both Methods

1. Subterranean termites (e.g., *Reticulitermes* sp.), Rotten-wood termites (*Zootermopsis angusticollis, Z. nevadensis*) or Dry-wood termites (e.g., *Kalotermes* sp.) should be used. For identification of termites see (9) and for identification of gut-protozoa see (2, 6, 7).
2. Keep termite species under optimal conditions (1).
3. Use only termites of the worker caste.
4. Identify live hind-gut microorganisms microscopically. Gut-smears stained with Heidenhains Hematoxylin may aid identification.
5. For microscopical inspection pull out the gut from the termite's posterior end with pointed tweezers. Open the gut in a drop of a salt solution. Neutral red may be added to the solution as a vital stain.

Method A. Details: For long-term tests, one may follow the testing method of the American Wood Preservers' Association (3). For short-term tests, petri dishes may be used for collecting fecal pellets. As a termite cellulose source filter paper may be used. To selectively eliminate hind-gut organisms, treat the filter paper with solutions of different concentrations of antibiotics and/or sulfanilamids. It may be necessary to try a wide selection of these compounds.

Gut examinations should be made at frequent intervals to determine the microorganism content of the gut. Determination of the test group feeding rates and the different utilization rates of the cellulose source (by fecal examination) helps to indicate the value of the eliminated gut organisms. The data obtained by comparing these results with the composition of the gut content would allow a better understanding of the host-symbiont relationship.

Method B. Details: *In vitro* cultures of the symbionts should be made under sterile conditions following basic microbiological methods. These cultures should be kept under the optimal host temperatures. Before inoculations of test tubes or petri dishes with gut contents are undertaken, the termites must be exteriorly sterilized by immersion for about 15 seconds each in an aqueous 1% mercuric-chloride solution, then a tincture of iodine solution, and finally in a 50% ethanol solution. To determine the nutritional and physiological requirements of the gut-organisms many different culture conditions should be tried. For example, determine whether aerobic or anaerobic conditions, which pH-range and which salts and other compounds benefit the growth of a culture. Determining the termite gut pH may indicate which pH-range to use for the cultures.

It is not difficult to get *in vitro* cultures of hind-gut bacteria. *In vitro* cultures of xylophagous hind-gut protozoa, however, are of more interest at this time. If xylophagous cultures can be obtained, then the quantity of cellulose metabolized should be determined. This could be accomplished by evaluating the deterioration of the given cellulose source or of the cellulase-activity in an *in vitro* culture. In addition, the investigator may attempt to make cellulase-activity measurements of the termite hind-gut.

Comparisons of the results obtained in Method A and B are of value in understanding the termite-symbiont system.

References · general
1. Krishna, K. and F. M. Weesner. 1969. Biology of termites. Vol. 1 and 2. Academic Press, New York and London.
2. Kudo, R. 1960. Protozoology. Charles C. Thomas, Springfield, Illinois.

· specific
3. American Wood Preservers' Association, Committee P-6. 1970. Methods for the evaluation of wood preservatives, Appendix A: Termite laboratory test method. Proceedings of AWPA, 66:167-170, revised 1971.
4. Cleveland, L. R. 1924. The physiological and symbiotic relationships between the intestinal protozoa of termites and their host with special reference to *Reticulitermes flavipes*. Biol. Bull. 46(4-5):178-227.
5. _____. 1925. Toxicity of oxygen for protozoa *in vivo* and *in vitro*: animals defaunated without injury. Biol. Bull. 48:455.
6. Kirby, H. 1946. Protozoa in termites. In Termites and termite control. 2d ed. C. A. Kofoid, ed. University of California Press, Berkeley.
7. Koidzumi, M. 1921. Studies on the intestinal Protozoa found in the termites of Japan. Parasitology 13:235-309.
8. Lund, A. E. 1960/61. Unpublished data. Personal communication.
9. Snyder, T. E. 1954. Order Isoptera. The termites of the United States and Canada. National Pest Control Assoc., New York.
10. Trager, W. 1934. The cultivation of a cellulose-digesting flagellate, *Trichomonas termopsidis,* and of certain other termite protozoa. Biol. Bull. 66(2):182-190.

20 / RELATIVE SENSITIVITY OF FUNGUS SPORES AND MYCELIUM TO TOXIC AGENTS

Ira M. Deep
Department of Plant Pathology
Ohio State University
Columbus, Ohio 43210

Malcolm E. Corden
Department of Botany and Plant Pathology
Oregon State University
Corvallis, Oregon 97330

Background

Tests for inhibition of spore germination and mycelial growth are run routinely in fungicide testing laboratories. A spore germination test is made by diluting the chemical under test to specific concentrations in a one percent sugar solution. A drop of the diluted chemical is then placed on a slide and a spore suspension is added. A spore can be considered to have germinated when the length of the germ tube exceeds half the diameter of the spore. Most fungal spores will germinate in 16-24 hours.

On making a mycelial inhibition test, known concentrations of the test chemical are included in a standard medium such as potato dextrose agar. The agar medium is prepared, sterilized, and cooled to 44° C before adding the test chemical. Then this special medium is poured into petri dishes, allowed to solidify, and finally inoculated with discs cut from old potato dextrose agar plates on which the test fungus has been growing.

It has been found that the mycelium of two fungi, *Fusarium oxysporum* f. *lycopersici* and *Fusarium roseum,* will grow on culture media containing 500 ppm of Phygon (2,3-dichloro-1,4-napthoquinone). Germination of spores from these two fungi can be completely inhibited by a concentration of Phygon below 5 ppm. Therefore, growth of the mycelium occurs at a concentration one hundred times greater than that needed to completely inhibit germination of spores.

Problem and Suggested Approach

This phenomenon may be unique to these two species of *Fusarium* or possibly spores may typically be more sensitive to the action of fungicides than the mycelium. Perhaps for some fungi, or with other chemicals, mycelial growth can be inhibited at concentrations that will not inhibit germination of the spores. Fungi are all around us and most of them produce spores. It would be a simple matter to collect 6 to 12 different fungi and to test the effectiveness of Phygon in inhibiting germination of spores and growth of the mycelium. The same tests could be run with other fungicides such as Captan and Fermate to determine whether this reaction is specific to Phygon. (Phygon is available from U. S. Rubber Co., Captan from Standard Oil Development Co., and Fermate from Du Pont Chemical Co.) (**Caution:** Remember that fungicides are poisonous substances, and care should be taken in working with them.)

It appears that certain metabolic reactions occur during germination of spores which do not occur during mycelial growth and that these reactions are inhibited by Phygon. How would one go about getting additional information about this phenomenon? Suppose the spores were allowed to produce well-defined germ tubes prior to exposing them to the chemical, would the germ tubes then grow as readily as the mycelium? If the spores were incubated for a period of time which was not long enough for germ tubes to form, but which allowed the beginning stages of germ tube development, would the germ tubes then continue to form even when exposed to the chemical? A distinction is herein made between reactions which bring about *initiation* of the germ tube and reactions which *regulate development* of the tube after its growth has begun.

Although the spores will not germinate in a 5 ppm Phygon solution, are they dead? If the spores were taken out of the Phygon solution and placed in dilute sugar water, would they then germinate? If so, the test chemical has acted as a fungistatic rather than a fungicidal substance.

The mycelium grows with a relatively high concentration of Phygon in the medium. Does the mycelium destroy some of the Phygon? This could be determined by culturing the mycelium in a Phygon solution, then determining whether the culture solution contained *less* than the original concentration of Phygon. Spore germination tests can be used as a biological assay to determine the concentration of Phygon in the solution. If all other factors are controlled, a given concentration of test chemical will cause a constant level of inhibition of spore germination. A standard curve can be prepared by determining the percentage spore germination in a series of concentrations of the test chemical. The Phygon solution in which a mycelial mat has been cultured should be compared with a Phygon solution of the original concentration.

Special Methods

These experiments require a minimum of equipment, but it would be necessary to prepare and maintain sterile media.

A pressure cooker can be used to sterilize media and glassware while an ordinary dissecting needle can serve as a transfer needle. When transferring cultures, the needle can be sterilized by flaming and cooled by dipping in 95% ethyl alcohol. Transfer work should be done in a closed room to avoid air currents, and precautions should be taken to avoid having dust particles from the air fall into the medium. You will find sources of information in the bibliography on the preparation of media and also on further techniques for making spore germination and mycelial inhibition tests.

References · general

1. Frobisher, M. 1957. Fundamentals of microbiology. 6th ed. W. B. Saunders Co., Philadelphia.
2. Horsfall, J. G. 1945. Fungicides and their action. The Ronald Press Co., New York.
3. Lilly, V. G. and H. L. Barnett. 1951. Physiology of the fungi. McGraw-Hill Book Co., New York.
4. Wilson, C. L. and W. E. Loomis. 1957. Botany. The Dryden Press, New York.

· specific

5. American Phytopathological Society, Committee on Standardization of Fungicide Tests. 1943. The slide-germination method of evaluating protectant fungicides. Phytopathology 33:367-632.
6. _____. 1947. Test tube dilution technique for use with slide-germination method of evaluating protectant fungicides. Phytopathology 37: 354-356.
7. Difco Laboratories, Inc. 1953. Difco manual of dehydrated culture media and reagents. 9th ed. Difco Laboratories, Inc., Detroit.
8. Sharvelle, E. G. 1961. The nature and uses of modern fungicides. Burgess Publishing Co., Minneapolis, Minn. (Chapter 16: Laboratory testing of fungicides).

21 / WOOD AND HUMIDITY*

Ramon Echenique-Manrique
Botany Department
Institute of Biology
National Autonomous University
of Mexico

Background

Wood is a hygroscopic material and, therefore, when there is a lot of humidity in the environment it tends to absorb it and it undergoes dimensional changes. The changes are of different magnitudes and can be produced in the three directions, longitudinal, radial, and tangential. The dimensional instability of wood is present only within certain margins of humidity content, that is, when the cell walls undergo changes in their moisture content. There are various methods that tend to give wood resistance to water absorption and dimensional stability.

Suggested Methods

What differences exist in the magnitude of the dimensional changes of wood in its three principal axes?

What is the principal cause for contraction or expansion in the longitudinal direction being different than the transverse directions?

What are the lower and upper limits of moisture content of wood that make it undergo dimensional changes?

What simple methods exist to reduce the tendency of wood to absorb moisture from the environment and to undergo swelling?

Suggested Methods

Prepare wood samples of oak or pine with a straight grain and free from defects in the following manner.

(a) Prepare a block 6 x 6 x 30 cm., in such a way that the annual rings appear as shown in the accompanying figure.

(b) Place the block in a container of water and keep it submerged for two weeks, changing the water every two days.

(c) From a saturated block cut five samples 5.0 x 5.0 x 0.5 cm., with the smaller dimension in the longitudinal direction of the fibers; prepare

*Edited on translation from Problemas de Investigacion en Botánica, Wiley S. A., 1971, with permission of the Consejo Nacional para la Enseñanza de la Biología.

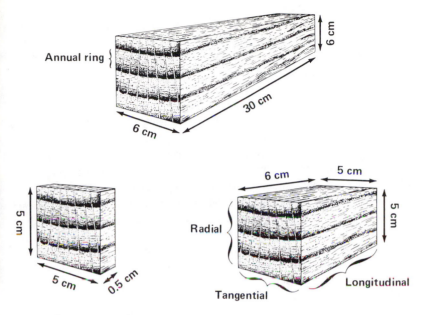

two samples 5.0 x 5.0 x 6.0 cm. with the longer dimension in the longitudinal direction of the fibers. During the entire cutting process you keep
the wood saturated with water and once the samples are prepared, they
should be returned immediately to the water container. The pieces should
be perfectly square with the dimensions indicated. (See figure.)

(d) Remove the pieces from the water container one by one and measure the tangential and radial dimensions of the small pieces and the longitudinal dimension of the large pieces with a micrometer (see figure).
Measurements should be accurate within .01 mm. Draw two diagonal lines
on the ends of the large pieces with a red pencil and measure the distance
from the point where the two lines cross to each side of the block of wood.
During these measurements you should make certain that the pieces of
wood remain impregnated with water. These dimensions will be called
"green."

(e) One by one dry the excess water from the surfaces of the pieces of
wood and weigh them to within .001 grams. These weights will be called
"green."

(f) Once the measurements and weights have been taken in the saturated or "green" state, you should let the pieces of wood dry in the open air. Place them on one of their sides in a place that is not exposed to conditions of dryness, such as air currents, rays of the sun, or high temperatures.

(g) Measure and weigh each piece again every 12 hours until you obtain a constant weight ($\pm$ .005 grams). The measuring and weighing operations should be made as close together as possible.

(h) Place the pieces of wood in an oven ($103 \pm 2°$ C) after they maintain a constant weight in the open air, and keep them there until they again maintain a constant weight of $\pm$.005 grams (approximately 36 hours). Take the pieces of wood from the oven and maintain them in a closed container (desiccator) with P_2O_5 under a screen. Leave them in a desiccator until they arrive at the environmental temperature. Measure and weigh the pieces of wood taking care that the time from when you take them from the desiccator to when you measure and weigh them is an absolute minimum. These values you will call "anhydrous."

(i) Calculate, using formulas 1 and 2, the moisture content (MC), the tangential contractions (TC), radial (RC), and longitudinal (LC) of the six samples used for each of the measurements.

Formula 1

$$MC = \frac{Wn - Wa}{Wa} \times 100$$

MC = moisture content (percentage)
Wn = the weight of wood in grams after it has arrived at a constant weight in the open air
Wa = the anhydrous weight in grams

Formula 2

$$C = \frac{Dn - Dg}{Dg} \times 100$$

C   = contraction (percentage)
Dn = normal dimensions in millimeters of the sample after it has stabilized in the open air
Dg = the "green" dimensions of the sample in millimeters

(j) On sheets of graph paper, one for each type of contraction, represent graphically the results obtained, where the X axis corresponds to the moisture content from 0 to the maximum obtained and the Y axis the percentage of contraction also from 0 to the maximum obtained. For the tangential and radial contraction on each sheet you will have five points for each measurement and two points for each longitudinal contraction.

Special Problems

At what moisture content do you begin to notice a change in dimension of the wood?

At what moisture content are the maximums and minimums of contraction? What type of contraction is maximal, minimal, and intermediate? What would happen to the dimensions of the anhydrous samples if these were saturated with water? What would be the configuration of a circular sample taken from its "green" condition and put into an oven until it arrives at its anhydrous condition? What should be the moisture content of wood before utilizing it in the fabrication of furniture so that the furniture, once it is in service, will undergo minimal dimensional changes?

You can increase the value of this problem by utilizing different types of wood to see the effect of the species of wood on the contractions of the wood. Also the pieces of wood can be finished with various varnishes, paints, waxes, etc., or impregnated with solutions of varnish and parafin dissolved in turpentine, to see what effects these treatments have on the dimensional changes. These treatments should be made before saturating the samples with water. In this case the saturation will be delayed more if you should begin by directly preparing the pieces of wood in the final dimensions, eliminating the preparation of the block 6 x 6 x 30 cm.

You can find general information on the structure of wood and its dimensional changes in references 2, 5, 6, 8, 11, 12, 13. There is data about methods of calculating dimensional changes in references 7, 8, 9, and in publications 1, 10, 13 the relationships between moisture content and dimensional changes of the wood are discussed. In Stamm's book (12) the efficiency of various methods of controlling dimensional changes is analyzed, and in bulletins 3 and 4 you can find data on dimensional changes in Mexican woods.

References

1. Anonymous. 1957. Shrinking and swelling of wood in use. U. S. Forest Service, Forest Products Laboratory Report No. 736. Madison, Wisconsin.
2. _____. 1960. Longitudinal shrinkage of wood. U. S. Forest Service, Forest Products Laboratory Report No. 1093. Madison, Wisconsin.
3. Echenique-Manrique, R. and Victor Diaz Gomez. 1969. Algunas características tecnólogicas de la madera de once especies mexicanas. Bol. Técn. 27:1-46. Instituto Nacional de Investigaciones Forestales, S.A.G. México, D. F.
4. Echenique-Manrique, R. 1970. Maderas Tropicales Mexicanas, serie Maderas de México. Cámara Nacional de la Industria de la Construcción, México, D. F.
5. The Forest Products Laboratory. 1955. Wood handbook. U. S. Department of Agriculture, Agriculture Handbook, No. 72:311-356. Washington, D. C.
6. Jane, F. W. 1954. The structure of wood. A.&C. Black Ltd. Londres.
7. McMillen, J. M. 1962. Methods of determining the moisture content of wood. U.S. Forest Service, Forest Products Laboratory, Report No. 1649. Madison, Wisconsin.
8. Panshin, A. J., Carl De Zeeuw, and H. P. Brown. 1964. Textbook of wood technology, Vol. 1. 2nd ed. McGraw-Hill Book Company, New York.
9. Peck, Edward C. 1960. Method of calculating shrinkage or swelling of wood with change in moisture content. U. S. Forest Service, Forest Products Laboratory, Report No. 1769-27. Madison, Wisconsin.
10. _____. 1961. Moisture content of wood in use. U. S. Forest Service, Forest Products Laboratory, Report No. 1665. Madison, Wisconsin.
11. _____. 1957. How wood shrinks and swells. Forest Prod. Journ. 7(7): 235-244. Madison, Wisconsin.
12. Stamm, Alfred J. 1964. Wood and cellulose science. The Ronald Press Company, New York.
13. Tarkow, H. 1960. Interaction of moisture and wood. U. S. Forest Service, Forest Products Laboratory, Report No. 2198. Madison, Wisconsin.

22 / EFFECTS OF AIR POLLUTANTS ON PLANTS*

Background

Raul Garza Chapa
University of Nuevo Leon
Monterrey, N. L.
Mexico

A large quantity of pollutants—solids, liquids, and gases are added to the air of the entire terrestrial globe. Among the principal sources of pollution are the internal combustion engine and the various industrial plants. There is no doubt that such pollutants are considerably affecting living beings.

Plants, wild or cultivated, are highly affected by air pollutants in different ways, such as in their development, anatomy, appearance, yield, and distribution. These effects cannot be disconnected from the interaction that is established among the contaminants and the biological and physical factors. Thus, we are obliged to know the ecology of plants.

The first organisms that are affected by atmospheric pollution are the plants. They can even be utilized as indicators, not only of the concentration but also the type of contaminant. A great variety of responses have been observed, and these depend on the various factors such as species, age, variety, plant leaf, environmental conditions (humidity, temperature, quantity of solar light present during the period of pollution, etc.), and others.

The problem of atmospheric pollution is serious and important; other than directly affecting man, it can cause serious damage in the plants to such a degree that forests, gardens, or agricultural zones could completely disappear. It is here that we must act as soon as possible.

Suggested Approach

The student can make observations under natural conditions of the vegetation that is found in the city (ornamental) and in the suburbs (wild and cultivated), or compare vegetation in an industrial center, where there is a large amount of automobile traffic, or in a small city, where traffic is light. The student can also make comparisons of the vegetation in different areas of the same city using such contrast factors as the presence of factories, type of factories, predominating air currents, etc.

---

*Edited on translation from Problemas de Investigacion en Botánica, Wiley S. A., 1971, with permission of the Consejo Nacional para la Enseñanza de la Biología.

It would be desirable to conduct a census of the species of predominant plants in the areas under comparison, for some plants can be reduced or eliminated by the effects of the pollutants. After that the student should compare the common species of the two areas in relation to their size and appearance. Look for characteristics such as the presence or absence of floration or some pathological symptoms such as necrosis and spots on the leaves, malformation of the organs, etc. If the student wishes to work in greater depth, he should look at the differences in the structures of the leaf tissues. He should be careful not to confuse the pathological symptoms with those caused by fungi, bacteria, virus, insects, and nutritive deficiencies.

At the same time the student should look for the differences in the pathological symptoms because these vary according to the type of pollutant that is present; this should be related to the nature of the source of pollution (type of factory, engines, etc.); among the most common are sulfur dioxide, floride, ethylene, ozone, and paraoxyacetylnitrate; these last two come from the action of light on the hydrocarbons and nitrogen oxides produced by internal combustion engines; in addition, there can be smoke, dust, and other particles.

Two types of investigations with controlled observations can be undertaken in open areas and in greenhouses. In both cases, the student can conduct studies on macroscopic and microscopic pathological symptoms (histological modifications); he can also conduct a chemical analysis of leaf tissues to see the concentration of chemical substances derived as a consequence of the pollutants.

In open areas it will be necessary to plant indicator species that detect the type of contaminant and the concentration of the contaminant. Among the plants that are possible to use, the following are recommended:

Pollutant	Plant
Sulfur dioxide	Alfalfa, dalia, sweet peas, poinsettia, and blackberry
Floride	Gladiolas, azaleas, tulips, young pines,
Ethylene	Orchids, carnations, tomatoes, cotton, blackeyed peas
Ozone and paraoxyacetyl-nitrate	Lettuce, petunia, tobacco, pinto bean, various vegetables

Under greenhouse conditions a great number of factors may be controlled. This includes the type of air, which must be purified by special types of filters (carbon is commonly used). It is also possible to vary the type of pollutant and its concentration.

With controlled factors there is a greater possibility of studying specific problems, such as special metabolic effects, histological and anatomical variations, concentration of pollutants in tissues, and variation among species in their response to pollutants.

References

1. Bobrov, R. A. 1952. The anatomical effects of air pollution on plants. Proc. Natl. Air Pollt. Sym., 2nd, Pasadena, California:129-134.
2. Brandt, C. S. and W. W. Heck. 1968. Effects of air pollution on vegetation. In Air pollution, A. C. Stern, ed. Academic Press, Inc., New York. 401-433.
3. Cole, G. A. 1959. Vegetation survey methods in air pollution studies. Agron. J. 50:553-555.
4. Darley, E. F. and J. T. Middleton. 1961. Carbon filter protect plants from damage by air pollution. Florist Rev. 127(3294):15-16, 43, 45.
5. _____. 1966. Problems of air pollution in plant pathology. Ann. Rev. Phytopathology 4:103-118.
6. Gordon, A. G. and E. Gorham. 1963. Ecological aspects of air pollution from an iron sintering plant at Wawa, Ontario, Can. J. Bot. 41:1063-1078.
7. Gucerian, E., H. Van Haut, and H. Stratmann. 1961. Problems of the recognition and evaluation of the effects of gaseous air impurities on vegetation. (Trans. from German by C. S. Brandt and P. Hotzel.) Tech. Report A-61-37, U. S. Dept. of H., E., W.
8. Heck, W. W. 1966. The use of plants as indicators of air pollution. Air Wat. Pollut. Int. J. 10:99-111.

23/ INTRASPECIFIC VARIATION IN PLANT PHENOLOGY

Elizabeth F. Gilbert
Department of Biology
Oberlin College
Oberlin, Ohio 44074

Background

The seasonal development of plants has been of interest to many individuals, more in the past than at present. Investigations in plant phenology have often been on a popular level, with information on the first wild flower blossoms appearing in the newspapers. Floral calendars have been made showing when one may expect to find local plants in bloom. More sophisticated work has endeavored to determine what factor or combination of factors of the environment control flowering. The progress of spring, as defined by the blossoming of certain plants, has been mapped for various regions of the world, and workers have attempted to frame laws governing the movement of flowering (i.e., "spring") across the country.

Suggested Problem

Very few quantitative data are available which can answer the question, What is the normal range of variation in the phenology of a certain species of plant? The date of first flowering is known more often than the date of flowering of the majority of individuals of one species. Other phenological events are even less well known than blossoming. The leafing-out and the appearance of the first fruit are of interest. Little is known of the time when the leaves or stems cease to grow or when the last flower may be found. Knowledge of the natural variation is basic to experimental work on the environmental control of phenological events, but information on the natural phenological variation within one species is still slight.

Suggested Approach

The greatest problem involved in studying phenology is the necessity for very frequent (perhaps daily) observations throughout the growing season or that part of the season in which a certain event occurs. Since observations may be made each day, the area of the study must be nearby. It may be desirable to identify individual plants by a system of tags; thus, the area should be relatively free from disturbance. Shrubs and herbaceous plants are easier to work with than trees. Records of observations should be kept along with relevant data on time and environmental conditions.

References · general

1. Daubenmire, R. F. 1947. Plants and environment. John Wiley & Sons, New York.
2. Odum, E. P. 1959. Fundamentals of ecology. W. B. Saunders Co., Philadelphia.
3. Weaver, J. E. and F. E. Clements. 1938. Plant ecology. McGraw-Hill Book Co., New York.

· specific

4. Caprio, J. M. 1957. Phenology of lilac bloom in Montana. Science 126 (3287):1344-1345.
5. Gilbert, E. F. 1961. Phenology of sumacs. Am. Midland Naturalist 66(2):286-300.
6. Huberman, M. A. 1941. Why phenology? J. Forestry 39:1007-1013.
7. Leopold, A. and S. E. Jones. 1947. A phenological record for Sauk and Dane Counties, Wisconsin, 1935-1945. Ecol. Monographs 17:81-122.
8. McMillan, C. 1957. Nature of the plant community. III. Flowering behavior within two grassland communities under reciprocal transplanting. Am. J. Botany 44:144-153.
9. _____ and B. F. Pagel. 1958. Phenological variation within a population of Symphoricarpos occidentalis. Ecology 39:766-770.

24 / OBSERVATION OF FLOWER POLLINATION BY ANIMALS

Robert L. Dressler
Smithsonian Tropical Research Institute
Panama, Canal Zone

Background

Pollination by animals is an important factor in the ecology of many species and has been a critical factor in the evolution and adaptation of many groups. Basically in this type of study one observes a flowering plant and ascertains which are the animals that arrive to visit the flowers. It would be desirable to make observations twenty-four hours a day during the entire flowering period of the species being studied, and in all parts of its geographical distribution, but this ideal is very difficult to attain. Here we offer a very generalized plan in the form of a "recipe" that can be adapted to the plants and environment where you may want to carry out studies of this type.

1. First, you must choose a species of plant. It is helpful if the plant is in its natural environment and if various plants of the species are accessible. For trees or lianas, simple observation may require the construction of platforms and ladders.

2. Before beginning observations, observe the development of the flower.

 a. Are the flowers open all day (and all night) or are they open only during certain hours?

 b. Do the flowers produce a perfume?

 c. Is the pollen accessible to insects at all times, or only during certain hours?

 d. Do the stigma and stamen mature simultaneously, or is the stigma receptive before or after the anthers shed their pollen?

 e. What appeared to be the principle attractants of the flower: color, perfume, pollen, nectar?

 f. What group(s) of animal(s) suggests the "syndrome" of the flower?

This will indicate to you at what time you must concentrate your observations and also how to carry them out.

3. If conditions permit, it would be desirable to investigate the genetic system of the plant.

    a. If you cover some recently opened flowers or buttons with cloth or other material that excludes possible pollinators, you can determine if the flowers can be autopollinated without the intervention of the pollinators. Remember that a very thick cloth or plastic can change the environment of the flowers and cause their death by excessive heat, humidity, or some other factor.

    b. Pollinate some flowers with pollen (1) of the same plant and (2) from other plants of the same species, making sure that these flowers will not be visited by any other pollinating agent. Thus, you can ascertain if the plant is autocompatible or not.

4. Observe the plant in bloom during all of its diurnal flowering cycle. If possible, attempt to observe the plant every five or ten days during its time of flowering in different sites.

    a. You must adjust your observation system to the animals that arrive. If the flowers are visited by very small insects, you will have to make your observations from a very close distance; however, an observer sitting at the side of a flower can scare away birds, bats, and even butterflies or bumblebees. If you must observe from some distance from the plant, it is recommended that you use binoculars.

    b. What animals visit the flowers?

    c. With what frequency do they visit the flowers?

    d. Observe the behavior of each species of animal that visits the flowers; which appear to pollinate the flowers? If it is possible, determine with certainty if the anthers place pollen on the body of the animal and if the same part of the body makes contact with the stigma.

5. Collect specimens of the plant and of each species of visiting animal. This is especially important because the majority of these studies are carried out either by botanists (who may not know the animals well) or by zoologists (who may not know the plants well). In case your observations are published, the specimens should be deposited in a museum or institute where future investigators can see the material.

1. Photography. Photographs of the pollination can be very useful and at times show details of behavior that have passed by unperceived. Many times it is preferable to place the camera on a tripod with the lens already focused on a flower or a group of flowers. This will permit you to take the photo with a cable without scaring the animal. With the use of flash, you can photograph the nocturnal visitors.

2. Red light. At night you can use red light without alarming the nocturnal animals that do not perceive it. A hand-held light covered by red cellophane is simple and efficient.

3. Study of the pollen carried by animals. Collect the animals and examine with a microscope the pollen that they carry on their bodies (examining the different parts of the body separately). The majority of the types of pollen can be used to identify the genus and even the species of the plant (Erdtman, 1954). In the case of bees, the females carry a quantity of pollen in the "pollen basket" (generally on the posterior tibia). Examining the pollen from the pollen basket, one can determine if the bees have been "constant" to a single species, or if they have collected pollen from various species. One can collect the bees in the flowers, or when they arrive at the beehive with their load of pollen.

4. Bait. For some groups of insects you can use a bait to attract many individuals. You can examine the insects to ascertain whether or not they carry pollen and identify the pollen. In the case of orchids and asclepiadaceae, a magnifying glass is sufficient to identify the pollen of these families.

    a. At night a light attracts many nocturnal butterflies. Entomologists sometimes utilize a "black light" with ultraviolet light that is more attractive for many insects.

    b. Fruit and something fermented, or a mixture of syrup, honey, and beer can attract many diurnal butterflies.

    c. Various aromatic substances such as eucalyptus, eugenol, vanilla, acetate, and the like attract male bees. These insects are important in the pollination of some orchids.

5. You can experiment by modifying the flowers or their environment and observing how these modifications affect the behavior of the pollinators.

POLLINATING ANIMAL	FLORATION	COLORS	PERFUME	FORM	POLLEN	NECTAR	OTHER CHARACTERISTICS
Coleoptera	Continuous	Dull or whitish	Strong, from fruit or rancid	Actinomorphous, shallow or forming a trap door	Abundant	Little or absent	Can have special nutritive bodies
Flies from carrion or excrement	Continuous	Dull, tan, chocolate, dark green, or polka-dotted	Strong and disagreeable	Actinomorphous, frequently forming a trap door	Little	Absent	They can have movable hair or other appendices. The trap door can have transparent "windows."
Bees	Diurnal or continuous	Brilliant, especially blue and yellow	Sweet	Highly variable, can be zygomorphous and deep	With abundant frequency	With abundant frequency	
Diurnal butterflies	Diurnal or continuous	Brilliant especially red and yellow	Weak	Tubular with a platform	Little	Abundant	
Nocturnal butterflies	Nocturnal	Pale, light blue or green	Strong and sweet	Tubular with or without a platform	Little	Abundant	Petals are sometimes outlined by a border.
Humming-bird	Diurnal	Brilliant especially red, yellow and lilac	Absent	Tubular without a platform	Little	Abundant	
Bats	Nocturnal	Subdued, light blue or green	Strong, rancid	Spherical inflorescence or a large and strong zygomorphous flower	Abundant	Abundant	Flowers, frequently on the trunk or drooping in large inflorescence.

Synthesis of some characteristics of flowers in relation to the group that pollinates them.

References

1. Alcorn, S. M., E. E. McGregor, and G. Olin. 1961. Pollination of saguaro cactus by doves, nectar-feeding bats, and honey bees. Science 133(3464):1594-1595.
2. Dodson, C. H., et al. 1969. Biologically active compounds in orchid fragances. Science 164(3885):1243-1249.
3. Dressler, R. L. 1968. Observations on orchids and Euglossine bees in Panama and Costa Rica. Rev. Biol. Trop. 15:143-183.
4. _____. 1968. Pollination by Euglossine bees. Evolution 22:202-210.
5. Erdtman, G. 1954. An introduction to pollen analysis. Chronica Botanica, Waltham.
6. Faegri, K. and L. Van Der Pijl. 1966. The principles of pollination ecology. Pergamon Press, Inc., New York.
7. Grant, V. and K. A. Grant. 1965. Flower pollination in the phlox family. Columbia Univ. Press, New York.
8. Macior, L. W. 1964. An experimental study of the floral ecology of *Dodecatheon meadia*. Am. J. Bot. 51:96-103.
9. Stephen, W. P., G. E. Bohart, and P. F. Torchio. 1969. The biological and external morphology of bees, with a synopsis of the genera of Northwestern America. Agric. Exp. Sta., Oregon State Univ., Corvallis, Oregon
10. Van Der Pijl, L. and C. H. Dodson. 1966. Orchid flowers, their pollination and evolution. Univ. of Miami Press, Coral Gables, Florida.

25 / HERBIVORES AND LIANAS

Background

Daniel H. Janzen
Department of Zoology
University of Michigan
Ann Arbor, Michigan 48104

In studying one of the group of climbing plants known as lianas, a botanist would look for the number of enlarged apexes, or tips, of the stems of the plant for much the same reason that a zoologist would look for the number of eggs in a bird's nest. Generally speaking, the more eggs in the nest, the higher the mortality rate of the young birds. Those that have a low mortality rate are those that grow slowly and that are given much protection by the adult. The same thing might be said of the lianas when the tip of the stem is very well protected and the plant does not depend upon a rapid growth. This can be observed, for example, in *Boraginaceae,* in some *Asclepiadaceae* (honeysuckle), and in some *Euphorbiaceae* (euphorbes). In such cases, only one or two enlarged stem tips are found.

By way of contrast, in birds that give their offspring relatively little protection, few of them survive. This is compensated for by the large number of offspring initially produced. Again, the same pattern is encountered in the lianas. In families such as *Convolvulaceae* and *Fabaceae,* used heavily as forage by plant-eating animals (herbivores), each root produces a large number of stem tips. Those that escape being eaten determine the plant's direction of growth.

Suggested Procedure

Studying the characteristics of lianas can be approached in two distinct manners. One way, is to sample the species of lianas in various types of habitats, recording the number of growing branches and the new points of growth for each group of roots; at the same time you would record the external proof of damage by insects or other animals. Such damage can take the form of eating the new growth of the branches, or the points of the stem. You should take into consideration in your sample the relative delicateness of the growth points and the presence of structures such as protective bracts, thick coverings, latex, or bitter compounds in the point of the same stem.

The second approach is more experimental in form. It is suggested that you choose four common species of lianas; two of them should have few points of growth and be well protected; the other two should have a large number of stem points, and these should appear to have been eaten by herbivores. You can choose fifty branches of new growth for each species, marking a point of reference on a convenient node or on the branch itself. The length of each one can be measured from the mark to the point, and you should record that length. After measuring the length, cut off all of the points of growth of half of the branches in the series with a pair of scissors. One week later you will proceed in the same manner and once again cut the points. Do not forget to record the number of points that you remove from each branch every time.

Continue to do the same thing for one or two months. The difference in response of the cut points of the stems in the experimental plants contrasted with the control plants should be quite evident. The information that this could give about the rate of growth of lianas would be useful and should be published.

Another Problem

The lianas can serve admirably in studying another example of the interaction between herbivores and plants. The basic information for this part of the study has never been obtained. What is needed is the amount of seed production for each species of liana in a particular area. The amounts can be measured in terms of:

1. the number of seeds;
2. the dry weight of the seeds;
3. the fresh weight of the seeds;
4. the individual size of each seed;
5. the individual weight of each seed;
6. the dimensions of the seed.

After taking these measurements (you should make between 10 and 20 different harvests of seeds), you will probably be impressed with the high rate of mortality of the seed because of the feeding activities of larvae, of beetles, and of butterflies. What we urgently need is an understanding of the relationship between the quantity of material that is gathered in the seed harvest contrasted with the quantity that is destroyed by the insects. The interaction between leguminous seeds and their predators has been discussed in some detail (3), but this was for a very restricted group of insects and a restricted group of plants. The lianas form many distinct families, and their seeds are eaten by many different insects. This diversity probably produces some characteristic relationships that were not evident in the study of the leguminous seeds and their predators.

One very important element in this study will be that of observing how the seeds from the mother plant are dispersed. The question is: Does it make any difference in the amount of seed eaten by insects whether the seeds are dispersed rapidly or slowly? It is probably that if they are dispersed rapidly, the insects will have little time to eat them.

The literature about the lianas is very scarce and is related principally, or most exclusively, to such statics as their morphology rather than to the dynamics of the individual or of the population. It would be desirable to consult with someone who is working actively with lianas, or with the depredation of their seeds, to obtain the additional information that may be necessary. This area of investigation is new and deserves much greater attention than has been given to it to date.

References

1. Janzen, D. H. 1966. Coevolution of mutualism between ants and acasias in Central America. Evolution 20:249-275.
2. _____. 1969. Allelopahy by mymecophytes: the ant Azteca as an allelopathic agent of Cecropia. Ecology 50:147-153.
3. _____. 1969. Seed-eater versus seed size, number, toxicity and dispersal. Evolution 23:1-27.

26/ VARIATION IN NATURALLY OCCURRING ANTIBODIES OF LABORATORY ANIMALS

Background

William H. Stone
Department of Genetics
University of Wisconsin
Madison, Wisconsin 53706

Antibodies are soluble proteins (gamma globulins) found in the blood serum of animals. The body manufactures them in response to contact with foreign substances called *antigens.* Smallpox vaccination is a familiar example of a deliberate injection of an antigen (the smallpox virus) to stimulate active antibody production and thereby make the vaccinated person immune or resistant to the disease. Recovery from certain diseases of bacterial or viral origin often makes the host resistant to further attacks by these organisms. This is because the natural infection, like the vaccination, causes the host to produce antibodies. Antibodies that are the result of some apparent immunization, such as vaccination or infection, are called *"immune"* antibodies.

But there is another kind of antibody called *"naturally occurring"* whose origin is unknown. Every animal contains some naturally occurring antibodies in its serum. The evidence indicates that immune and naturally occuring antibodies are not actually physically or chemically different. However, there are two schools of thought as to the origin of naturally occurring antibodies. One proposes that they are the result of some unknown infection or invasion of antigen and thus are really immune antibodies. The other thinks that their production is genetically determined and thus they are manufactured by the body as a regular function just as blood cells or other body tissues. It may be that both of these proposals are correct.

Biologists have studied naturally occurring antibodies since the turn of the century, when it was discovered that they were very important in human blood transfusions. Since the serum of one person may contain naturally occurring antibodies capable of reacting with the antigens of another person's blood cells, serious consequences may result if incompatible blood is transfused. Therefore, blood typing of both donor and recipient is routinely practiced by hospitals to avoid transfusion accidents.

A few years ago, it was discovered in cattle that a naturally occurring antibody, called anti-J, reactive with cattle red blood cells, showed what appeared to be a "seasonal" variation in its concentration. The concentration was determined by testing increasing dilutions of the serum with a standard panel of positively reacting red cells. The highest dilution of the serum showing reactivity is designated as the *titer* of the serum. Samples of sera obtained repeatedly over a period of time from a single cow showed that the anti-J antibodies varied considerably in titer, and that this variation had a definite pattern. The most interesting discovery was that the variation appeared to be associated with the season of the year. For example, serum samples obtained from August to October had relatively high titers, whereas samples obtained from December to March had relatively low titers. This same kind of fluctuation in titer was found in a large number of cattle.

A similar fluctuation was discovered later in the anti-A and anti-B antibodies of humans, the same antibodies that are of great concern in blood transfusions. The naturally occurring anti-R antibodies of sheep also showed this same seasonal variation. Evidence that this variation is truly associated with season, at least in cattle, was obtained from a study in South Africa. The rise and fall of anti-J titer in South Africa (southern hemisphere) showed a six-month difference as compared to the time of rise and fall in anti-J observed in the United States (northern hemisphere).

Definition of the Problem

Although the phenomenon of seasonal variation of naturally occurring antibodies seems well established for anti-J of cattle, anti-A and anti-B of humans, and anti-R of sheep, it is not known whether all naturally occurring antibodies exhibit this phenomenon. Furthermore, there is no information available on the physiologic basis of this phenomenon. There are reports of a variety of environmental factors that influence antibody production, but none of these is concerned with naturally occurring antibodies. For example, it is known that climatologic phenomena have a direct bearing on certain physiologic activities, but to what extent these activities influence naturally occurring antibodies is unknown.

It would be desirable to look for this phenomenon in small laboratory animals such as rabbits, rats, mice, guinea pigs, or hamsters. Any one or more of these species that exhibit the phenomenon could then be used to study the factors that may influence the variation in naturally occurring antibodies. For example, if it could be established that the naturally occurring anti-A (human) known to occur in rabbit normal sera showed seasonal variation, a number of adult rabbits could be subjected to different environmental conditions (dark vs. light, hot vs. cold, etc.) and the effect on the antibody levels could be studied. Samples of sera could be obtained from the rabbits at various intervals and assayed for their antibody content. The procedure for assaying antibodies in serum is available from any of several texts on immunology, or can be obtained gratis from drug firms who supply reagents and equipment to blood banks.

Supplies and Equipment

It is necessary to have suitable facilities to house and care for a number of laboratory animals. The facilities should allow for some variety of environmental conditions, particularly light and temperature.

Various serologic equipment such as needles, syringes, test tubes, pipettes, incubators, and racks should be available. The only large laboratory equipment necessary is a centrifuge, a refrigerator and freezer or combination. Human blood can usually be obtained gratis from local blood banks.

References · general

1. Dunsford, I. and C. C. Bowley. 1956. Techniques in blood grouping. Charles C. Thomas, Springfield, Ill.
2. Hildemann, W. H. 1970. Immunogenetics. Holden-Day, San Francisco, California.
3. Race, R. R. and R. Sanger. 1968. Blood groups in man. 5th ed. F.A. Davis, Philadelphia, PA.

· specific

4. Leigh, Egbert Giles, Jr. 1970. Natural selection and mutability. Am. Natur. 104(937):301-305.
5. Lewin, B. M. 1970. The molecular basis of gene expression. John Wiley & Sons, New York.
6. Rendel, J. 1957. Further studies on some antigenic characters of sheep blood determined by epistatic action of genes. Acta. Agr. Scand. 8: 162-190.

7. Seligy, V. L. and J. M. Neelin. 1970. Transcription properties of stepmise acid-extracted chicken erythrocyte chromatin. Biochem. Biophys. Acta. 213(2):380-390.

8. Shaw, D. H. and W. H. Stone. 1958. "Seasonal" variation of naturally occurring iso-antibodies of man. Trans. VI Congr. European Soc. Hematol. S. Krager, New York.

9. Stone, W. H. 1956. The J substance of cattle. III: Seasonal variation of the naturally occurring iso-antibodies for the J substance. J. Immunol. 77:369-376.

10. _____. 1962. The J substance of cattle. Annals N. Y. Acad. Sci. 97: 269-280.

11. _____. 1962. The relation of human and cattle blood groups. Transfusion 2:172-177.

12. _____. 1967. Immunogenetics of type-specific antigens in animals. In Advances in immunogenetics by Tibon J. Greenwalt. Lippincott, Philadelphia, PA.

27 / SLOW LACTOSE FERMENTATION BY BACTERIA

Irving P. Crawford
Department of Microbiology
Western Reserve University Medical School
Cleveland, Ohio 44101

Background

An aspect of microbial genetics that has received a great deal of attention recently is the control of the structure of proteins such as enzymes. For each enzyme one or more genes exist that control the unique amino acid sequence characteristic of that enzyme's structure.

Many laboratories are engaged in studies where mutations are induced in bacteria to provide altered enzymes. It is likely that many such mutations have occurred naturally, however, and that some of the resulting bacterial strains, forming altered enzyme molecules, have survived for lack of selective pressure against them. Examples of this might be found in the inducible enzyme β-galactosidase. An inducible enzyme is one that is produced in the presence of a specific substrate (substance acted upon) or a substance that is structurally similar to the substrate, but cannot be detected when the cells are grown in media free of the inducing substance. In the present case, bacterial cells will produce β-galactosidase only in the presence of lactose or a structurally similar inducer. This enzyme permits the utilization of lactose as a carbon and energy source by causing the hydrolysis of lactose into the constituent simple sugars, glucose and galactose. The presence of an altered form of β-galactosidase would probably not confer a great disadvantage on a bacterial strain unless it were required to grow on lactose.

The property of lactose utilization (usually ascertained as the ability to *ferment* lactose, that is, to produce acids and/or gases from it anaerobically) has been used for many years by bacteriologists as an aid to identification of unknown strains. There exists a vast amount of information in this field of microbiology that might yield interesting results if examined genetically.

Suggested Approach

Nonpathogenic members of the *Enterobacteriaceae* are suggested as prime subjects for investigation, both because there are so many of them with impaired ability to ferment lactose, and because the normal mechanism of lactose utilization has been well studied, genetically and biochemically, in certain members of this group.

It should be possible to obtain strains of slow lactose fermenters from any diagnostic bacteriological laboratory. The diagnostician would be likely to classify them either in the genus *Salmonella* or as members of the paracolon group. **Caution:** Careful bacteriological techniques must be used in handling these bacteria. Some so-called nonpathogenic bacteria may, under certain circumstances, become pathogenic. Perhaps better than working with an unknown strain would be the acquisition of known slow lactose fermenters from the American Type Culture Collection. These could then be tested to determine whether or not they did, in fact, ferment lactose appreciably slower than they could ferment the constituent simple sugars. For example, in the usual fermentation tests a slow lactose-fermenting strain might show definite evidence of fermenting glucose or galactose within 18-24 hours, but may not show evidence of fermenting lactose for many days.

Having determined that the culture as a whole is impaired in its ability to ferment lactose, the first of many questions can be asked. Do *all* the cells of your stain ferment lactose *slowly* or is there a mixed population of *good* lactose fermenters and cells *unable* to form β-galactosidase in the culture? With a little thought, an experiment to distinguish the alternatives can be devised remembering (1) that β-galactosidase is usually an inducible enzyme, (2) that a medium can be devised where lactose utilization is *necessary* for growth as well as one where it is inconsequential, and (3) that inoculum size can be decreased to the point where any tube or culture showing growth most probably resulted from the introduction of one viable cell. These experiments alone would comprise a worthwhile project, and they could be accomplished without elaborate equipment.

If in the preceding experiments a culture is found where all of the cells ferment lactose slowly, other questions arise that might require more complicated techniques for their elucidation. Is the enzyme produced normal in function but low in amount (as if the induction mechanism were faulty), or alternatively, is the cell unable to produce β-galactosidase as rapidly as it can other enzymes? Is the enzyme simply inefficient in hydrolyzing lactose? Could there be a mixture of enzyme molecules produced, a few normal ones and many inactive ones? Does lactose enter the cell readily? Obviously at this point the investigation could take several directions, the most promising ones probably requiring quantitative estimation of the β-galactosidase activity present per cell, per milligram of protein in a cell extract, and perhaps per immunological unit (measured with anti-β-galactosidase serum). Other approaches may suggest themselves. Certainly the influence of inducer (substrate) concentrations should be investigated.

This project would not, of course, be an easy one, and it is impossible to foresee all the pitfalls. If attacked, however, it would give the investigator an idea of the nature of some of the problems currently facing microbiologists. If successfully pursued, the investigation might yield information and strains that could later be used either in biochemical and genetic investigations of gene action, or in immunogenetical work.

References · general

1. Bracken, A. 1955. The chemistry of microorganisms. Pitman Publishing Corp., New York.
2. Conn, H. J. and M. W. Jennison, eds. 1957. Manual of microbiological methods. McGraw-Hill Book Co., New York.
3. Gunsalus, I. C. and R. Y. Stanier, eds. 1961. Bacteria: a treatise on structure and function. Vol. 2. Metabolism. Academic Press, New York.
4. Lederberg, J. 1950. Isolation and characterization of biochemical mutants of bacteria. Methods in Medical Research 3:5-22.
5. Stacey, M. and S. A. Barker. 1960. Polysaccharides of microorganisms. Oxford University Press, New York.
6. Stainer, R. Y., M. Doudoroff, and E. A. Adelberg. 1962. The microbial world. 2d ed. Prentice-Hall, Inc., Englewood Cliffs, N. J.

· specific

7. Ciferri, Orio, et al. 1970. Uptake of synthetic polynucleotides by competent cells of *Bacillus subtilis*. J. of Bact. 104(2):684-688.

8. Cohn, M. 1957. Contributions of studies on the β-galactosidase of *Escherichia coli* to our understanding of enzyme synthesis. Bacteriol. Rev. 21:140.

9. Hoch, S. O., et al. 1971. Control of tryptophan biosynthesis by the methyltryptophan resistance gene in *Bacillus subtilis*. J. of Bact. 105 (1):38-45.

10. Novick, A. and M. Weiner. 1959. The kinetics of β-galactosidase induction. Chapter 6 in R. E. Zirkle; ed., Symposium on molecular biology. University of Chicago Press, Chicago.

11. Schafler, S., L. Mintzer, and C. Schafler. 1960. Acquisition of lactose fermenting properties by salmonellae. II. Role of the medium. J. Bacteriol. 179:203.

28 / ANTIMICROBIAL SUBSTANCES FROM SEEDS

Background

W. E. Grundy
Abbott Laboratories
North Chicago, Ill. 60064

Microorganisms, especially the actinomycetes, produce a variety of antibiotics. As a result, they have been studied very extensively, especially by industrial research laboratories. Although higher plants are known to produce antimicrobial substances, much less work has been done on them. Tomatine is an example of such a compound. This material is found in the leaves of the tomato plant and several other *Lycopersicum* species. The compound has been crystallized and the chemical structure is known. Although Tomatine has not become a marketable antibiotic, it has been tested against human fungus infections of the skin and found to have activity. Undoubtedly other interesting compounds exist and will be found.

Most of the work with plants has been concerned with the leaves, stems, roots, fruit, or bark. Relatively little effort has been devoted to the seeds. A few investigators have examined seeds and their data indicate that seeds may be a good source of antimicrobial substances. Some varieties of seeds are resistant to mold growth in germination tests while other seeds are highly susceptible. This is additional evidence that seeds contain substances inhibitory to microorganisms.

Suggested Approach

This study could take any one of several directions. If a large variety of seeds is collected, a general survey for the presence of antimicrobial substances could be carried out. This might indicate whether the presence of such substances is widespread in seeds, or whether they are restricted only to certain types of seeds. As described in Reference 6, such a survey can be made using simple techniques, without extracting the antimicrobial substances from the seeds. Perhaps a more intensive study can be made of one particular substance found in one variety of seed. The antimicrobial spectrum of the substance could be determined by testing its effectiveness against various types of bacteria, yeasts, and molds.

For one who is more chemically minded, there is the challenge of extracting, isolating, and purifying the antimicrobial agent. To do this, it would be necessary to work out a method of extraction which does not destroy the substance. Try solvents such as chloroform, ethanol, dilute acid, dilute base, neutral salt solution, acetone. It might be worthwhile to attempt several extraction procedures, and finally select the one which proves to be most effective.

To study a particular antimicrobial substance in detail, an assay procedure must be developed. This involves the selection of a rapidly growing, yet sensitive organism. The well-known paper disc technique provides a quantitative method of assaying the potency of the substance, since the zone of inhibition can be measured rather readily. The dilution method of assaying potency is an alternate approach.

Comments on Difficulties and Equipment Needs

Solvents such as chloroform or alcohol and acids or bases have antimicrobial properties. You must be careful not to confuse this effect with the substances extracted from seeds. The solvents may be evaporated from the extracts and the residues tested for activity. Acidic or basic solutions may be neutralized before they are tested. Antimicrobial agents are often rather unstable compounds, and treatment with heat or with strong acids and bases should be avoided.

You should be familiar with the basic bacteriological techniques such as the preparation of culture media, sterilization, culture transferring and maintenance, and aseptic technique. Equipment such as petri dishes, bacteriological test tubes, Erlenmeyer flasks, a sterilizer or pressure cooker, and a bunsen burner or alcohol burner are required.

The equipment required for the extraction of antimicrobial agents from seeds would include such items as mortar and pestle, beakers, funnels, filters, pipettes, and volumetric cylinders. The extraction can be done with simple equipment. If chemical glassware is readily available you can develop rather elaborate setups. It is well to bear in mind that since these extracts will be tested bacteriologically they should be as free from contaminating microorganisms as possible.

If you decide to study the antimicrobial agent chemically it would be advisable to have the advice of a chemist in planning the experiments, especially in purifying and concentrating the active substance.

References · general
1. Gause, G. F. 1960. Search for new antibiotics. Yale University Press, New Haven, Conn.
2. Grove, D. C. and W. A. Randall. 1955. Assay methods of antibiotics. Medical Encyclopedia, New York.
3. Wren, R. C. 1956. Potter's new cyclopedia of botanical drugs and preparations. Pitman Publishing Corp., New York.

· specific
4. Carlson, H. J. and H. G. Douglas. 1948. Screening methods for determining antibiotic activity of higher plants. J. Bacteriol. 55:235-240.
5. _____, and J. Robertson. 1948. Antibacterial substances separated from plants. J. Bacteriol. 55:241-248.
6. Ferenczy, L. 1956. Antibacterial substances in seeds. Nature 178:639-640.
7. Malekzadeh, F. 1966. Antibacterial substance from cauliflower seeds. Phytopathology 56:497-501.
8. Maruzzella, J. C. and M. Freundlich. 1959. Antimicrobial substances from seeds. J. Am. Pharm. Assoc. Sci. Ed. 48:356-358.

29 / ACQUIRED IMMUNITY IN PLANT VIRUS DISEASES

John R. Keller
Department of Biology
Seton Hall University
South Orange, New Jersey 07079

Background

A major advance in virology has been the recognition that the presence of one virus strain in living tissue can protect against infection by another strain of the same virus. This idea has had more widespread application in animal virology, with the development of vaccines, than in plant virology, but the principle is the same in both cases. Almost all viruses, such as polio virus and tobacco mosaic virus, comprise different strains. Some strains may produce a mild reaction in the host; some may produce a severe reaction; others may actually kill the host.

It has been established that once an infective virus particle lodges in a cell and begins multiplication, a particle of another strain of the same virus cannot become established in that cell. A person infected with a mild strain of polio virus throughout the body tissues has immunity to a severe or fatal strain of polio. Similarly, a plant systematically infected with a mild strain of tobacco ringspot is protected against a lethal strain of the same virus. Animals have a natural defense against "foreign" proteins such as viruses by the production of antibodies in the blood. This can provide a long-lasting immunity from recurrent virus infection even though the virus itself may have disappeared from the body. Essentially this is the mechanism of immunity.

Plants do not form antibodies. Protection of a plant by one strain of virus against another exists only as long as there is active virus multiplication taking place in the cells. However, the principles involved in plant virology and the study of strain protection with tobacco ringspot virus strains or tobacco mosaic virus strains provide a valuable and challenging approach for the study of the mechanism of immunization.

Suggested Problem
There are several possible methods for studying acquired immunity (often called cross-protection) in plants. One may establish a systemic infection of a mild virus strain in a plant and then after a period of time inoculate the new growth of the plant with the severe strain. The new growth may then be observed to see whether it is immune to the severe strain.

A better method utilizes the "half-leaf technique," where a local lesion reaction occurs; here a quantitative study can be made. See Bawden (1) and Smith (4). One can also observe how thoroughly the mild strain becomes established systemically, how long it takes, and the effect of age of the leaf on the severity of the reaction.

Try to obtain a plant virus that can be worked with easily. Such a virus (a) produces definite symptoms that easily identify its presence, (b) has several strains, and (c) is common enough to be obtained readily. Good possibilities for study are tobacco mosaic and potato virus X strains. Both have mild and severe strains so that protection by the mild strain against the severe strain can be tested. Also, if one keeps a careful observation on a mild strain such as that of tobacco mosaic virus, one may find yellow flecks in the leaf which indicate the appearance of a mutation to a severe strain.

A question that is often raised in plant virology is whether it may be feasible to inoculate fruit trees with a mild virus strain for protection. There is a danger that the mild strain can mutate to a severe one. There is also the problem of a new infection by a different virus. The effect of the two viruses may prove fatal to the host.

Equipment and Procedure

The equipment needed for this problem should not be too difficult to obtain. Tobacco plants can be grown easily in a greenhouse, if the school has one. Otherwise, a group or bank of incandescent lights or fluorescent lamps can be suspended over an area so that plants can be grown by artificial illumination. Clay pots and soil provide the best situation for growing plants individually so that one can move the plants about as desired.

The tobacco varieties necessary are *Nicotiana tabacum* var. turkish and var. Samsun; *N. glutinosa* can be used rather than the Samsun variety. If one is using potato virus X then *Gomphrena globosa* should be grown. Type virus strains can be obtained from the American type culture collection, 12301 Parklawn Drive, Rockville, Maryland 20852. If one is working with tobacco mosaic virus strains PV 42, tobacco mosaic masked strain M, and PV 39 tobacco mosaic yellow aucuba strain are suggested.

The principal consideration is to get strains of the same virus that can be distinguished from each other by the host plant reaction. Also, the virus should be able to give a reaction in a local lesion host called a "test plant." The student should be familiar with the "test plant" for the particular virus he or she has in mind to work with.

The techniques for preparing inoculum and inoculating plants are reasonably simple. The simple technique of using a cheesecloth swab with carborundum dust on the plant has been well described. See Smith (4) and Bawden (1) on inoculations. It is necessary to have adequate controls. These should consist of uninoculated plants and plants that have been rubbed with a cheesecloth swab using water in place of virus inoculum. This latter method is used especially in the half-leaf technique.

Plants inoculated with a mild strain can be checked for virus multiplication by taking a portion of a leaf and preparing from it an inoculum that is introduced to the test plants. Under certain environmental conditions, such as higher temperature, infected tobacco plants appear uninfected (masking effect), whereas in reality they are diseased. For this reason, periodic sampling of plants infected with the mild strain must be carried out as a test for virus multiplication.

If the half-leaf technique is employed, count the local lesions and assemble the data in some statistical form. Read material on live polio virus vaccine for an application to humans.

References · general

1. Bawden, F. C. 1964. Plant viruses and virus diseases. 4th ed. The Ronald Press Co., New York.
2. Crafts, A. S. 1961. Translocation in plants. Holt, Rinehart & Winston, New York. pp. 28-33. (movement of viruses).
3. Matthews, R. E. F. Plant virology. Academic Press, New York.
4. Smith, K. M. 1951. Recent advances in the study of plant viruses. 2d ed. McGraw-Hill Book Co., New York.
5. _____. 1960. Plant viruses. 3d ed. John Wiley & Sons, New York. pp. 107-160.

6. Stanley, W. M. and E. G. Valens. 1961. Viruses and the nature of life. E. P. Dutton and Co., Inc., New York.

· specific

7. Bennett, C. W. 1953. Interactions between viruses and virus strains. Advances in Virus Research 1:30-67.

8. Hitchborn, J. H. and A. D. Thomson. 1960. Variation in plant viruses. Advances in Virus Research 7:163-187.

9. Knight, C. A. 1959. Variation and its chemical correlates. In F. M. Burnet and W. M. Stanley, eds., The viruses, Vol. 2. Academic Press, New York. pp. 127-156.

10. Kunkel, L. O. 1934. Studies on acquired immunity with tobacco and Acuba mosaics. Phytopathology 24:437-466.

11. ———. 1955. Cross protection between strains of yellow-type viruses. Advances in Virus Research 3:251-273.

12. Matthews, R. E. F. 1949b. Studies on potato virus X. II. Criteria of relationships between strains. Ann. Appl. Biol. 36:460-474.

13. McKinney, H. H. 1929. Mosaic diseases in the Canary Islands, West Africa and Gibralter. J. Agr. Res. 29:557-578.

14. Price, W. C. 1932. Acquired immunity to ringspot in Nicotiana. Contrib. Boyce Thompson Inst. 4:349-403.

15. ———. 1936a. Virus concentrations in relation to acquired immunity from tobacco ringspot. Phytopathology 26:503-529.

16. ———. 1936b. Specificity of acquired immunity from tobacco ringspot diseases. Phytopathology 26:665-675.

17. Salaman, R. N. 1933. Protective inoculation against a plant virus. Nature 131:468.

18. Siegel, A. and S. G. Wildman. 1954. Some natural relationships among strains of tobacco mosaic virus. Phytopathology 44:277-282.

· References for Polio Virus

1. Koprowski, H., G. A. Jervis, and T. W. Norton. 1952. Immune response in human volunteers upon oral administration of a rodent-adapted strain of poleomylitis virus. American J. of Hygiene. 55:108-126.

2. Sabin, A. B. 1959. Present position of immunization against poliomyelitis with live virus vaccines. British Medical Journal 1:663-680.

3. Salk, J. 1959. Poliomyelitis: control No. 24. p. 499. Viral and Rickettsial diseases of man. 3rd ed. Lippincott Co., Philadelphia.

30 / STUDIES ON NATURAL POPULATIONS

Background

Murvel E. Annan
111 Highland Avenue
Staten Island, New York 10314

In nature various genetic and environmental factors affect the number of individuals of a species that populates a given locality. Among these individuals are some which are heterozygous for a given trait and others which are homozygous for the same trait.

It has often been said that in sexually reproducing organisms, such as the vinegar fly, *Drosophila*, genetic heterozygosity is favored by natural selection over genetic homozygosity. That is, individuals heterozygous for gene alleles will leave more offspring than homozygous individuals. It is difficult to demonstrate just how heterozygous individuals in nature really are, but there are some indirect methods by which the importance of heterozygosity could be determined. One obvious method is to lower the level of heterozygosity of a population and see what effect this has on the ability of the more homozygous individuals to leave offspring.

Suggested Problem

The ability to leave offspring is a complex character. In *Drosophila*, for example, it is compounded of egg-laying ability, hatchability of the eggs, survival of hatched larvae to adulthood, longevity of adults and so on. A demonstration that any of these characters is affected by the level of heterozygosity would be an important contribution to our understanding of the genetic structure of natural populations.

Suggested Approach

Flies of the genus *Drosophila* are suitable organisms for this kind of study. The ease with which the flies can be collected, the availability of many standard strains, and the comparatively simple equipment necessary for setting up an adequate laboratory are factors contributing to the suitability of these flies for this sort of work. Directions for collecting the flies can be found in Demerec (3). Many available standard strains are listed in Reference 5. The setting up of a laboratory for *Drosophila* research is described in Demerec and Kaufmann (4) and Hovanitz (5). Descriptions of the common species of *Drosophila* are assembled in the volume edited by Patterson (7). Changing the heterozygosity of flies can be accomplished by inbreeding. Full-sib (brother-sister) mating is the most effective way, in *Drosophila*, of increasing homozygosity. In each successive generation of full-sib mating a greater proportion of gene loci is made homozygous. (For the relationship between number of generations of full-sib mating and heterozygosity see Chapter 19 of Sinnott, Dunn, and Dobzhansky [1].) It is not difficult to make full-sib pair matings for several successive generations while making observations.

Individual wild females, inseminated in nature, can be placed in individual containers such as ¾-oz. creamers. Each such female is then the original parent of a "line." When offspring hatch, several pairs of brothers and sisters are placed individually in creamers to produce the next generation. Thus, in each "line" there will be several "sublines." These sublines are necessary in order not to lose a whole line due to failure of a single pair to produce offspring. Generation after generation, a single pair is mated from each subline to produce the next generation.

As a control line a mass culture of about 20 males and 20 females can be maintained without brother-sister mating.

A good character to measure each generation is egg production. The entire lifetime egg production of a female can be measured by putting the female on fresh food daily or every two days and then counting the eggs laid. Hatchability of these eggs can be easily determined if they are left for a day or two and then the number of empty and full egg cases counted. Longevity of males and females is another good character to measure and will come out of the measurements of lifetime egg production without additional work. Sample techniques for these procedures are described in Reference 6.

Possible Pitfalls

If sib mating results in loss of fertility then the lines will be harder and harder to keep going as time goes on. An adequate number of sublines should be started (perhaps 5 to 10 per line) to allow the experiment to go on for at least 10 generations without dying out.

Another difficulty to anticipate is the likelihood of considerable variation between individual flies in performance. This increases the number of individuals that it is necessary to observe in order to obtain valid results.

References · general

1. Sinnott, E. W., L. C. Dunn, and T. G. Dobzhansky. 1958. Principles of genetics. 5th ed. McGraw-Hill Book Co., New York.

· specific

2. Carl, Ernest. 1971. Population control in Arctic ground squirrels. Ecology 52(3):395-413.
3. Demerec, M., ed. 1950. Biology of *Drosophila*. John Wiley & Sons, New York.
4. _____ and B. P. Kaufmann. 1961. *Drosophila* guide. Carnegie Institution of Washington, Washington, D. C.
5. Hovanitz, W. 1953. Textbook of genetics. D. Van Nostrand Co., Princeton, N. J.
6. Novitski, E., arr. 1962. *Drosophila* information service (36). Prepared at the Department of Biology, University of Oregon, Eugene.
7. Patterson, J. T., ed. 1943. Studies in the genetics of *Drosophila*. Part 3. The Drosophilidae of the Southwest. No. 4313. University of Texas Press, Austin.
8. Tamarin, R. H. and S. R. Malecha. 1971. The population biology of Hawaiian rodents: demographic parameters. Ecology 52(3):383-394.

31/ THE ROLE OF COMPETITION IN DETERMINING THE INTENSITY OF NATURAL SELECTION

Background

Richard C. Lewontin
Department of Biology
University of Chicago
Chicago, Illinois 60637

The Darwinian theory of evolution states that some genotypes in a population of a given organism have a higher reproductive efficiency than others. As a result of this greater reproduction, the favored type will increase in frequency; possibly it will replace the less favored type.

There are two factors that strongly influence the reproductive efficiency (fitness) of a genotype. One is the physical environment—temperature, humidity, food material, among others. The other factor is the number and kinds of other organisms interacting with the genotype being investigated. In some cases two genotypes may compete directly with each other for some resource (water, food) in short supply. In other cases two genotypes may actually facilitate each other; the fitness of each may be increased by the interaction between them. There is already some experimental evidence on the effect of the interaction of different genotypes in determining fitness (3,5,6,7,8, and 9). It is, however, restricted to a few cases and considerably more evidence is needed. Evidence of this sort would be of great importance in understanding one factor affecting evolution in a population.

Problem

The problem is to determine in what way the presence of different genotypes in different proportions in a population affects the fitness of the genotypes to survive.

A good organism to use for such a problem is *Drosophila melanogaster*, the fruit fly. *Drosophila* is easily cultured (4) and its genetics is well known. One component of fitness may be studied—the survival of individuals from egg to adult. This would be a measure of competition mainly in the larval stage when feeding and growth occur.

The basic technique is simply this:

1. Collect eggs in the following way. Spread a small amount of *Drosophila* food (4) on a milk bottle cap. Put one or more mated females of a single genotype into the bottle, and invert the bottle onto the cap. After the eggs have been laid, discard the females. Examination of the food under the dissection microscope will enable you to count out the number of eggs you want to transfer to a culture vial.

2. Place a known number of eggs on a determined amount of food—perhaps a culture vial with ½ inch of standard *Drosophila* food. A sufficiently large number of eggs should be used so that not all the larvae succeed in reaching adulthood. The number of eggs could be 40, 60, 80, 100—depending on the size of the container.

3. Keep track of the total number of larvae which hatch from the eggs in each vial. This constitutes a check on the viability of the eggs as well as on the suitability of the medium and of the environmental conditions under which your experiment is being conducted.

4. Count and record the number of adults which eventually hatch out. Each of these adults represents a larva which was successful in completing its life cycle to adulthood. The figures you get will enable you to calculate the percentage of success for each genotype in each specific competitive situation.

In carrying out the actual experiment, known proportions of eggs of different genotypes may be placed in the same culture vial so that the effect on survival of different mixtures can be determined. By using strains homozygous for recessive mutations, or heterozygous for dominant mutants the adult flies can be recognized and classified, and a determination can be made of the proportion of larvae of each genotype which completed the life cycle.

A specific experiment may be designed as below. Naturally, this would have to be one of a series of similar experiments. Assume that each container will contain 60 eggs, and that we are using mutants of wing form.

Vial No.	Wild type	Vestigial Wing	Dumpy Wing	This will give a measure of
1	60 eggs			survival rate of each type when not in competition with other types
2		60 eggs		
3			60 eggs	
4	30 eggs	30 eggs		survival rate of one type when in competition with one other type.
5	30 eggs		30 eggs	
6		30 eggs	30 eggs	
7	20 eggs	20 eggs	20 eggs	survival rate in competition with two other types.

Once the survival rate without competition has been determined for each genotype, this can be used as a base against which to compare survival rates in populations of varying genotypic proportions.

You can increase or decrease the proportions of different genotypes and see if changes in initial population alter the survival rates.

If you plot a graph of your results similar to the hypothetical one below, you can come to some conclusions concerning survival rates for each competitive situation. Naturally, you should not expect to find a curve as smooth and regular as those in the illustration.

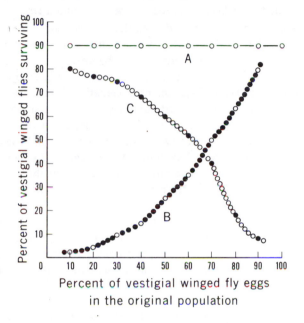

A line such as A would imply that a given percentage of vestigial winged flies will survive regardless of its proportion in the original population. A curve like B would imply that as the proportion of vestigial winged fly larvae in the population is increased, its ability to survive also increases. Curve C would indicate the converse.

A second problem to be attacked might be the effect of other *species* on the relative survival of *D. melanogaster* genotypes. For this purpose various combinations of *D. melanogaster* mutants with a wild strain of *D. simulans*, a closely related species, may be used. As in the previous case, eggs would be mixed in known proportions. Percent survival of the different genotypes and the two species might then be observed.

1. Before you begin such an experiment you must perfect a technique for culturing and handling *Drosophila*. Demerec and Kaufmann (4) is an excellent reference which outlines all the basic procedures. Start by rearing cultures of wild type flies, and when you have mastered the technique, expand your culture collection to include various mutant strains.

Starter cultures can be obtained from the Department of Genetics, Cold Spring Harbor, New York 11724, if your teacher writes on school stationery. Often the genetics department of a nearby university will give you the flies you need to get started. A complete collection of *Drosophila* is kept at the University of Texas, Department of Genetics, Austin, Texas 78712, and starter cultures can be obtained upon payment of a nominal fee.

2. A list of mutant strains from which you might select your experimental material is given in Demerec and Kaufmann (4, page 33). In selecting the strains to be used in the actual experiment, consider the following factors:

a. Ease of identification. In a mixed population it might be difficult to recognize a fly that differs from the wild type by a single bristle somewhere on the abdomen.

b. Quantitative traits. Traits that overlap should not be used. For example, Bar eye is quite viable, but if it is used in a population mixed with eyeless where the eye is reduced to one quarter to one half normal size, it may be difficult to distinguish members of the two genotypes from one another, although with practice these can be unambiguously distinguished.

c. Is the trait recognizable in the larva? A trait such as yellow body might be exceptionally good for this type of study, since larvae of this genotype can be identified by the yellow-brown color of the mouthparts.

3. It is already known that many heterozygotes are stronger and can survive better than homozygotes (3). You must be sure that greater survival of one genotype does not result from greater heterozygosity in that strain. You can avoid this pitfall to some degree by inbreeding each mutant strain for a number of generations before beginning your experiments. A method of reducing heterozygosity by inbreeding is described by Annan in the previous problem.

4. Remember that in this experiment you are determining the fitness of certain larvae to live in competition with larvae of other genotypes. You

are not testing the fitness of adults of one genotype to compete with adults of another genotype. Therefore, a trait affecting wing structure may have little or no effect in the particular competition you are testing. On the other hand, the genotype for wing structure may be associated with larval characteristics that have a definite relationship to larval survival. We do not know. This is what your investigation might reveal after it has been completed.

References · general

1. Sinnott, E. W., L. C. Dunn, and T. G. Dobzhansky. 1958. Principles of genetics. 5th ed. McGraw-Hill Book Co., New York.
2. Snyder, L. H. and P. R. David. 1957. The principles of heredity. 5th ed. D. C. Heath & Co., Boston.

· specific

3. da Cunha, A. B. 1949. Genetic analysis of polymorphism of color pattern in *Drosophila polymorpha*. Evolution 3:239-251.
4. Demerec, M. and B. P. Kaufmann. 1961. *Drosophila* guide. Carnegie Institution of Washington, Washington, D. C.
5. Lewontin, R. C. 1955. The effects of population density and composition on viability in *Drosophila melanogaster*. Evolution 9:27-41.
6. _____ and Y. Matsuo. 1963. Interaction of genotypes determining viability in *Drosophila busckii*. Proc. Natl. Acad. Sci. 49:270-278.
7. Merrell, D. J. 1951. Interspecific competition between *Drosophila funebris* and *Drosophila melanogaster*. Am. Naturalist 85:159-169.
8. _____. 1953. Selective mating as a cause of gene frequency changes in laboratory populations of *D. melanogaster*. Evolution 7:287-296.
9. Moore, J. A. 1952. Competition between *D. melanogaster* and *D. simulans*, I. Evolution 6:407-420.
10. _____. 1952. Competition between *D. melanogaster* and *D. simulans*, II. Proc. Natl. Acad. Sci. 38:813-817.

32 / CYTOLOGICAL EVIDENCE BEARING ON THE PHYLOGENETIC RELATIONSHIP BETWEEN *ANEMONE* L. AND *HEPATICA* MILLER

Charles R. Freitag
6100 Inwood Street
Hyattsville, Maryland 20734

Background

It is sometimes difficult to determine the degree of relationship that exists between organisms that possess certain common characteristics. Are there enough similarities to include the organisms in the same genus—the same species? In addition to a study of morphological characteristics, careful examination of the chromosomes of such organisms often reveals information leading to a decision concerning degree of relationship. Thus, cytological analysis may help scientists to improve the taxonomy (classification) of organisms.

In recent years the genus *Hepatica* Miller has been generally treated as separate from the genus *Anemone* L., yet the separation is not accepted by all botanists. Evidence bearing particularly on the degree of homology displayed by chromosomes of representative species of the two genera might help considerably in determining the suitability of this separation. This kind of evidence would be a product primarily of successful hybridization. It appears that no attempts have been made to cross any species of *Anemone* with any of *Hepatica*. Should such crossing be successful, it would make possible (through determination of the degree of homology exhibited by the chromosomes of meiotic microsporocytes) some sort of conclusion as to the probable degree of. relationship between the genera. Should it prove unsuccessful, the failure would at least lend support to the proposition that the two genera should remain separate. Evidence pointing in either direction would represent a real contribution to the sum of our knowledge.

Suggested Approach

First familiarize yourself with the local flora and the locations where specimens of *Anemone* and *Hepatica* may be found. It is desirable to start the project in the late summer or early autumn, while the plants are in vegetative condition and specimens may be collected and potted. They should be left outdoors but protected (for example, under a layer of leaves) until February or March, when they should be brought indoors. Access to greenhouse space is desirable, but almost any sunny indoor area will do. Bringing the plants indoors at this early date will expedite blooming and some time can thus be gained.

During the winter months when the plants are dormant, familiarize yourself with the relevant literature. Arrange to obtain or have access to the necessary equipment. After the plants have been brought indoors, set aside a certain number of them for use in preliminary studies. These studies should include: (1) collecting stamens at various short intervals of time, making temporary aceto-orcein smear stain mounts, and examining the mounts for meiotic microsporocytes; (2) noting how long it takes the flower buds to reach the stage at which the microsporocytes are actively undergoing meiotic division; (3) determining the simplest satisfactory means of effecting cross-pollination of species at the same time safeguarding against random pollination; (4) making slides of aceto-orcein stained smears of a few root-tips, study of which may aid in understanding and interpreting the microsporocyte chromosomal configurations seen in the stamen smears.

During the period of preliminary studies it will also be necessary to do the actual cross-pollinating of a number of specimens of each species with those of the other species. The flowers should be protected until the achenes (fruits) start to ripen. The mature dried achenes should be carefully saved and stored in labeled containers. Unless specific sources can be located that will describe methods for germinating the seed of species of *Anemone* or *Hepatica*, it is best to sow the seed in the autumn on the surface of sterile potted soil, the soil being as much as possible like that in which the parent plants were found. The pots should then be covered with dead leaves and allowed to remain outdoors all winter in a protected location.

The following spring, they should be examined periodically to see if any seed has germinated. If so, the plants should be handled as much as possible as if they were growing wild. When (and if) the hybrids bloom, prepare stamen smears and examine them to find microsporocytes undergoing meiotic division. The degree to which chromosomes are paired or partially paired in meiotic microsporocytes should be an indication of the closeness of the relationship between the particular species crossed. In the interest of making available to other workers as much of the basic data (that is, the appearance of the chromosomes) as possible, include in your research report an extensive series of *camera lucida* drawings of the chromosome configurations found in the meiotic microsporocytes.

Special Considerations

1. The two species of *Hepatica* found in North America occur only between the east coast and about 95° west longitude, and between about 30° and 46° north latitude.

2. Care will have to be used to select only spring flowering species of *Anemone*.

3. There is no way of knowing for sure whether or not the hybridization will produce viable seed, or how long any seedlings produced will take to come into flower. The evidence points to the probability that species of *Hepatica* require several years to come into bloom, but it is possible that this might be hurried by artificial methods of temperature and light control.

4. There is need for a good microscope equipped with an oil immersion lens, mechanical stage, and condenser; a *camera lucida* attachment; a few chemicals (for example, acetic acid, orcein stain, 95% and 100% alcohol); a small amount of laboratory glassware; and other equipment, including a small dissecting kit, abosrbent cotton, label tags, and gelatin capsules.

5. The project might also be carried out with other species available locally, although the choice of species to be used would probably require the advice of someone familiar both with cytotaxonomic and related literature and with the local flora.

References · general

1. Baker, J. R. 1958. Principles of biological microtechnique. John Wiley & Sons, New York.
2. Biological Abstracts. Published by University of Pennsylvania, Philadelphia.
3. Elliott, F. C. 1958. Plant breeding and cytogenetics. McGraw-Hill Book Co., New York.

4. Gray, P. 1958. Handbook of basic microtechnique. 2d ed. McGraw-Hill Book Co., New York.

5. Gurr, E. 1956. A practical manual of medical and biological staining techniques. 2d ed. Interscience Publishers, New York.

6. Laurie, A., D. C. Kiplinger, and K. S. Nelson. 1958. Commercial flower forcing. 6th ed. McGraw-Hill Book Co., New York.

7. Mahlstede, J. P. and E. S. Haber. 1957. Plant propagation. John Wiley & Sons, New York.

8. Northen, H. T. and R. T. Northen. 1956. The complete book of greenhouse gardening. The Ronald Press Co., New York.

9. Sharp, L. W. 1943. Fundamentals of cytology. McGraw-Hill Book Co., New York.

10. Sinnott, E. W., L. C. Dunn, and T. G. Dobzhansky. 1958. Principles of genetics. 5th ed. McGraw-Hill Book Co., New York.

11. Snyder, L. H. and P. R. David. 1957. The principles of heredity. 5th ed. D. C. Heath & Co., Boston.

12. Swanson, C. P. 1957. Cytology and cytogenetics. Prentice-Hall, Inc., Englewood Cliffs, N. J.

· specific

13. Bresch, C. and R. Hausmann. 1970. Classical and molecular genetics. Translated. Springer-Verlag: Berlin. New York, U.S.A.

14. Darlington, C. D. 1956. Chromosome botany. The Macmillan Co., New York.

15. _____ and A. P. Wylie. 1956. Chromosome atlas of flowering plants. 2d ed. The Macmillan Co., New York.

16. Demerec, M., ed. 1956. Advances in genetics. Vol. 8. Academic Press, New York.

17. Eberle, Paul. 1970. Chromosomal criteria for the selection of experimental subjects. Translated. Med. Monatsschr 24(8):341-346.

18. Gregory, W. C. 1941. Phylogenetic and cytological studies in the Ranunculaceae Juss. Trans. Am. Phil. Soc., new series, 31(5):443-521.

19. Johnson, L. P. V. 1945. A rapid squash technique for stem and root tips. Canadian J. Research 23(C):127-130.

20. LaCour, L. 1941. Aceto-orcein: a new stain-fixative for chromosomes. Stain Tech. 16:169-174.

21. Lewis, Kenneth R. and J. Bernard. 1970. The organization of heredity. Am. Elsevier Publ. Co.

22. Matsuura, H. and T. Sutō. 1935. Contributions to the idiogram study in phanerogamous plants. J. Fac. Sci. Hokkaido Imperial Univ., series five, 5:33-75.

23. Moffett, A. A. 1932. Chromosome studies in *Anemone* L. Cytologia 4:26-37.

24. Steyermark, Julian A. and Cora S. Steyermark. 1960. *Hepatica* in North America. Rhodora 62:223-232.

25. Wells, B. and L. F. LaCour. 1971. A technique for studying one and the same section of a cell in sequence with the light and electron microscope. J. Microscopy 93(1):43-48.

33 / INTRASOMATIC SELECTION OF RADIATION-INDUCED MUTATIONS IN THE TOMATO

Background

Thomas R. Mertens
Department of Biology
Ball State University
Muncie, Indiana 47306

Ever since the late 1920's, when Muller and Stadler demonstrated that X-radiation induces mutations, it has been recognized that most of the mutations induced are harmful in that they do not promote the survival of the species under ordinary conditions. Occasionally, however, evidence has been presented to show that in plants, at least, radiation may induce beneficial mutations. One technique for screening such beneficial mutations was used with some success in the tomato (4). This technique was based on the hypothesis that if tomato seeds were irradiated and mutations harmful to plant growth and development were induced, the cells that contain these mutations would be selected *against* in the development of the plant. Conversely, cells containing induced mutations beneficial to growth and development would be selected *for* as the plant develops. Consequently, flowers developing on the second or third inflorescence (flower-cluster) should contain fewer detrimental and more beneficial induced mutations than the flowers of the first inflorescence.

Although there is some evidence for such "intrasomatic selection," much of it is based on casual observation. For example, in seedlings grown from irradiated tomato seeds, a high frequency of chlorophyll sectoring has been noticed in the first true leaf, but a much lower frequency in the second and subsequent leaves. The concept of intrasomatic selection would be placed on much sounder ground if there were some accurate quantitative data to support it.

Two recent research notes (6 and 7) by van der Mey and De Nettancourt indicate that some progress has been made on the problem of intrasomatic selection in the tomato. These Dutch workers have studied the elimination of radiation induced chromosome aberrations in the course of root tip development. They have also employed pollen inviability as a means of tracing cell lines in the development of inflorescences of tomato plants. To date, however, no data have been published on variation in mutation frequency as a function of the inflorescence examined.

Suggested Approach
Tomato seeds of an autodiploid or highly inbred line of tomatoes could be X-rayed. Relatively few irradiation doses need be used, but they should be spread over a wide range, for example, 5,000 rontgens, 10,000 r, 20,000 r, 40,000 r. Additional unirradiated control plants should be grown concurrently. It would be desirable to have twenty-five plants grown to maturity at each level of irradiation. The greater the number of plants grown from irradiated seed, the greater the likelihood of obtaining plants with radiation-induced mutations. The seeds could be planted and grown to maturity either in the field or in a greenhouse.

Three types of data might be gathered from these plants to test the hypothesis of intrasomatic selection: (1) Pollen could be removed from anthers of the flowers on each of the first three inflorescences and sampled for viability. This can be done by staining the pollen with aniline blue. Viable pollen will stain well; defective grains stain poorly (*see* Johansen). If radiation-induced chromosome aberrations are selected against, pollen viability should be greater in inflorescences 2 and 3 than in the first inflorescence. (2) Microsporogenesis in flowers from each of the first three inflorescences might be studied. The incidence of chromosome aberrations— translocations, inversions, and so forth—should be lower in the upper inflorescences than in the first if intrasomatic selection occurs. (3) Flowers on each of the first three inflorescences could be self-pollinated and the resulting seeds planted and checked for mutations in the seedlings. One would expect three normal plants to every one mutant if a recessive mutation were induced in the embryo of an irradiated seed. In many respects this information on mutation would be the most significant evidence concerning intrasomatic selection. If intrasomatic selection eliminates detrimental mutations, one would expect to observe a lower mutation rate in the second and/or third inflorescence than in the first.

Problems and Pitfalls

1. One of the first problems in executing this work is to obtain a suitable inbred or autodiploid line of tomatoes. Although it is not absolutely necessary to do so, using such a near-homozygous line will insure greater uniformity in the parent line (control) and enable one to attribute any variation to the radiation treatment. These lines of tomatoes are available

through the Tomato Genetics Cooperative. Dr. C. M. Rick of the Department of Vegetable Crops, University of California Davis, could provide information on sources for such seed. (It would be desirable to request information on school stationery and with the teacher's approval.)

2. A second problem is that of irradiating the seeds. There are a number of possible solutions to this problem: a local hospital, college or state university may have facilities to X-ray seeds. In fact, some national laboratories may give assistance; Argonne, Brookhaven, or Oak Ridge might provide the needed facilities.

3. Careful, detailed record-keeping is essential. If done well, this research will be time-consuming and laborious. This, no doubt, accounts for the present scarcity of detailed observations on intrasomatic selection.

References · general

1. Hollaender, A. ed. 1954-56. Radiation biology. 3 Vols. McGraw-Hill Book Co., New York.
2. Johansen, D. A. 1940. Plant microtechnique. McGraw-Hill Book Co., New York.
3. Strickberger, M. W. 1968. Genetics. The Macmillan Co., New York.

· specific

4. Mertens, T. R. and A. B. Burdick. 1957. On the X-ray production of "desirable" mutations in quantitative traits. Am. J. Botany 44:391-394.
5. _____ and R. R. Gomes. 1956. Phenotypic stability in rate of maturation of heterozygotes for induced chlorophyll mutations in tomato. Genetics 41:791-803.
6. van der Mey, J. A. M. and D. De Nettancourt. 1970. Somatic elimination in tomato plants irradiated with fast neutrons at the dry seed stage and at the seedling stage. Report of the Tomato Genetics Cooperative 20:27-28.
7. _____. 1970. Pollen abortion as a criterion of chimerism in the sporogenic tissue of the M_1 plant after seed irradiation. Report of the Tomato Genetics Cooperative 20:28-29.

34 / A COMPARATIVE STUDY OF THE PECTORAL GIRDLE IN THE HYLIDAE

Richard J. Baldauf
Department of Wildlife Science
Texas A & M University
College Station, Texas 77840

Background

The study of evolution is the most general and, in a sense, most important branch of biological endeavor, for it is the study of the past, present, and future of all of life. It involves an understanding of the natural relationships among organisms and of the relations of the organisms with their environment. All biological knowledge is grist for the mill of evolutionary study. There are some fields, however, that are of special importance if we are to understand the phylogeny and adaptations of plants and animals through millions of years of time. Chief among these is morphology, for the structure of organisms is preserved in fossils and living species alike, giving valuable clues to their relationships and their modes of adaptation to their environment.

Despite its great importance, morphology has occupied a "back seat" for many years, and this is understandable. The basis for most systems of classification had been worked out by those famous zoologists of the Golden Age of Morphology during the nineteenth century. What followed was a burst of profound thinking in allied, newer fields of genetics, serology, physiology, and ecology as applied to speciation, hybridization and, therefore, evolution and systematics.

However, one cannot dispose of morphology so easily. First, a thorough understanding of vertebrate phylogeny depends heavily on fossils, and these must, of necessity, be described and understood according to the structures preserved. Unfortunately the fossils cannot be subjected to genetical and physiological experimentation. Thus modern paleontologists spend much of their time studying the morphology of today's vertebrates in efforts to better understand the structure of fossils.

Second, the morphology of an organism is not simply an accident, but represents the result of a long evolution of successive adaptations of living things to their environment. Seen in this perspective, the structures of modern vertebrates present some extremely interesting problems.

One such problem in phylogeny and adaptation can be studied in various families of frogs and toads. In particular, the closely related families Bufonidae, Leptodactylidae, and Hylidae all have arciferal pectoral girdles, but the structure of the pectoral girdles is not the same in all these families. It is not even the same in all the members of the Hylidae, or tree frogs, but shows specializations for habitat, digging, and time of emergence from hibernation. Perhaps there is also a correlation between the structure of the girdle and amplexus.

A study of the variations, differences, and similarities in the spatial relationship and size of the bones and cartilages composing the pectoral girdle in tree frogs will contribute to our knowledge of the structure, evolution, and activities of these animals.

Suggested Approach

Several adult specimens of various species of the family Hylidae must be obtained by field collecting or by requesting gifts of excess specimens from large museums and universities. Since the cartilages must be included in the study, it will be necessary to study *cleared specimens* rather than dried skeletons (in which the cartilages are often lost). Special directions for preparing cleared specimens are given.

The study of the pectoral girdle of cleared specimens should include considerations of size, position, and extent of ossification of the pectoral elements. These considerations may, in certain cases, be expressed mathematically to show points of stress and strain in relation to movements of the anterior appendages. In this regard the pectoral girdle offers a remarkable opportunity to relate trigonometry to functional morphology.

The similarities and differences noted in the structure of the pectoral girdles might possibly lead to a "grouping" of the species included in the study. Such a "grouping" would constitute a tentative scheme of classification of the species studied and might well lead to basic facts necessary to understanding the relationships among the tree frogs. This, in turn, will contribute to our knowledge of the general evolution of the Hylidae, particularly since a detailed study, as proposed here, would indicate characteristics invaluable to the study of fossil hylids.

Procedure

1. Become familiar with the pectoral girdle of frogs by consulting the references.

2. Prepare the specimens for study by clearing.

3. With a dissecting microscope, study the pectoral girdle of each species.

4. Write a detailed description of the pectoral girdle for each species, with additional notes on the structural variations between specimens of the same species. These descriptions should include relative size and position of pectoral bones and cartilages for each species.

5. Prepare drawings to show the characters described in (4) above.

6. Compare the descriptions of the species studied by noting differences and similarities.

7. Group the species, if possible, according to similarities in structure.

8. Read pertinent literature on habits of the species studied to determine possible correlations between structure and habits. (Example: A more robust pectoral girdle might be correlated with the early appearance of the species after a minimum of spring rains. Species with a weak girdle might be able to dig out of the hibernation site only after the ground is thoroughly softened by spring rains.)

A Simplified Process for Clearing and Staining Small Vertebrates

The following procedure, based on a process described by Davis and Gore, is modified to suit a few selected vertebrates. By this process it is possible to render specimens transparent and to stain the bony parts of the skeleton. The end result is a beautiful preparation. Mice, frogs and toads, and lizards are suggested as suitable animals for the procedure as given here. Other animals can be prepared, but with some difficulty and often with the need for special chemicals and techniques.

The procedure is not difficult, but there are pitfalls; these will be recognized as you proceed. Do not become discouraged if the first few specimens are spoiled. It's like baking a cake—you will soon learn when you need an extra pinch of a certain ingredient or a little extra time for a particular step. These uncertainties are, for the most part, due to the variation in specimen size. A larger specimen will normally require a little more in time and chemicals than a smaller one.

I. Preparing the Animal

1. Kill by placing in a bottle of 60% ethyl alcohol. This will leave the animal in a relaxed condition.

2. Remove from alcohol immediately after death. With scissors, cut a longitudinal slit in the belly to expose the intestines. Immerse specimen in a bottle or pan of 10% formalin, prepared as follows:

9 parts of water

1 part of commercial (full strength) formalin (usually expressed as 37% to 40% formaldehyde by weight)

Keep in 10% formalin for about one week. A longer time, however, will not spoil the specimen.

3. Skin the specimen. Using tweezers and scissors, go through the abdominal slit and remove the viscera. Also remove each eyeball by cutting the muscles that attach it to the socket. Remember, you are preparing this specimen to display bones and cartilage, so do not cut or break any part of the skeleton.

II. Staining after Preliminary Clearing

1. Put the specimen in a jar containing 4% potassium hydroxide (KOH) and place the jar by a window for sunlight (for bleaching). To prepare the 4% KOH, dissolve 4 grams of dry KOH in 100 cc of water. The amount of the KOH solution should be twice the volume of the specimen; that is, if the specimen displaces 50 cc of water, use 100 cc of 4% KOH. This same rule applies to all following steps.

The KOH will cause the muscles to become partially transparent. Keep the specimen in 4% KOH until you can see the outline of the femur. For an average-sized frog this step will take three or four days. The specimen will fall apart if kept too long in KOH, and since other steps require its use you should move from step to step on a reasonable schedule.

2. Do not transfer the specimen. Instead, pour out the 4% KOH and replace with fresh 4% KOH. To this add, dropwise, 1 cc of stain for each 100 cc of 4% KOH. This is not a critical amount; enough stain should be added to turn the KOH a deep wine color.

The stain is prepared as follows:

Glacial acetic acid	5 cc	Chloral hydrate (1% solution)	60 cc
Glycerine	10 cc	Alizarin Red S (dry powder)	

The amount of powdered stain is determined by adding it slowly, while shaking the liquid mixture, until no more powder will dissolve.

Keep the specimen in stain until the bones are colored (look at the femur again). The process of staining should take no more than 24 hours, perhaps less. It is important to remove the stain as soon as the bones are colored; otherwise, the soft tissues will pick up too much stain (a little is all right).

3. Pour out the stain and add fresh 4% KOH. This will remove excess stain from the specimen and will give you a better chance to check on the bones. If the bones appear to need more stain, pour out the KOH and replace with stain solution. Don't forget, you are looking at a partially cleared specimen; some bones may appear to be understained because they are deeply imbedded in muscles.

Keep in 4% KOH for about 24 hours.

III. Clearing the Specimen

1. Pour out the 4% KOH and add Clearing Solution No. 1:

Glycerine	20 parts
4% KOH	3 parts
Water	77 parts

Keep in No. 1 solution for about two or three days. This will partially clear the tissues.

2. Pour out Solution No. 1 and add Clearing Solution No. 2:

Glycerine	50 parts
4% KOH	3 parts
Water	47 parts

Keep in No. 2 for two or three days, or longer if the tissues continue to become more transparent.

3. Pour out the No. 2 solution and add Clearing Solution No. 3:

Glycerine	75 parts
Water	25 parts

Same comments as in No. 2 above.

4. Pour out the No. 3 solution and add pure glycerine. The specimen should reach its maximum transparency during this step. Store the specimen in the same glycerine. A growth of mold may appear on the surface of the glycerine in time; this can be prevented by adding a crystal of thymol.

Note: Clearing Solutions 1, 2, and 3 may be reused. The extent of reuse depends on the size of specimens. Store solutions for reuse in separate bottles.

Precautions

Formalin will corrode metals. Keep it stored in a glass bottle with a plastic or glass-rubber gasket lid. Formalin and its fumes are irritating to the eyes and other mucous membranes. Wash it off your hands after use.

KOH is a strong alkali. Do not get it onto mucous membranes. Again wash your hands after use. KOH will bleach and eat through fabrics; do not let it splash.

Chloral hydrate is a narcotic (knockout drops). Store this chemical in a locked cabinet; it should not be available to children.

STUDENTS SHOULD USE THESE THREE CHEMICALS ONLY UNDER THE DIRECT SUPERVISION OF A TEACHER.

References · general

1. Noble, G. K. 1954. The biology of the Amphiba. Dover Publications, New York.

· specific

2. Cochran, D. M. and C. J. Goin. 1970. New field book of reptiles and amphibians. G. P. Putman's Sons, New York.
3. Cope, E. D. 1866. On the structure and distribution of the genera of the arciferous Anura. J. Acad. Nat. Sci. (Philadelphia), 6:67-112.
4. Davis, D. D. and V. R. Gore. 1936. Clearing and staining skeletons of small vertebrates. Field Museum Nat. Hist. Tech. Series (4):1-15.
5. Goin, C. J. and O. B. Goin. 1971. Frogs and toads. *In* Introduction to herpetology. 2d ed. W. H. Freeman & Co., San Francisco.
6. Hoffman, A. C. 1930. Description of the external characters and of the shoulder girdle of *Heleophryne*; and a note on the systematic position of the genus. South African J. Sci. 17:414-425.
7. Howell, A. B. 1935. Morphogenesis of the shoulder architecture. Part 3. Amphibia. Quart. Rev. Biol. 10:397-431.
8. Hsiao, S. D. 1933. A comparative study of the pectoral region of

some typical Chinese Salientia. Peking Nat. Hist. Bull. 8:169-204.

9. Jones, E. J. 1933. Observations on the pectoral musculature of Amphibia Salientia. Ann. Mag. Nat. Hist. 12:403-420.

10. Noble, G. K. 1926. The pectoral girdle of the brachycephaliid frogs. Am. Mus. Nov. (230):1-14.

11. Parker, W. K. 1868. Structure and development of the shoulder girdle and sternum in the Vertebrata. Ray Soc. (London) 1-238.

12. Procter, J. B. 1921. On the variation of the scapula in the Batrachian groups Aglossa and Arcifera. Proc. Zool. Soc. (London) (1):197-214.

13. Romer, A. S. 1924. Pectoral limb musculature and shoulder-girdle structure in fish and tetrapods. Anat. Record 17:119-143.

14. Savage, J. M. and S. B. Emerson. 1970. Central American frogs allied to Eleutherodactylus bransfordii (Cope): a problem of polymorphism. Copeia 4:623-644.

35 / VARIATION IN MOUTHPARTS OF TADPOLES

Background

Arthur N. Bragg
Professor of Zoology
University of Oklahoma
Norman, Oklahoma 73069

The mouthparts of amphibian tadpoles have long been used as an aid in identification of species and subspecies. Usually two opposing tendencies occur in the pattern of these mouthparts: (1) a general similarity within any one taxonomic unit or species, as opposed to (2) considerable variation in detail among individuals of the same species or subspecies. The extent of variation in each species should be more fully investigated, and, in particular, the causes of the variation (heredity, food supply, temperature or other factors) established. For no species is the range of variation fully known; the descriptions of many kinds of tadpoles have been based upon too few specimens to show even the possible extent of the variations and almost nothing as to their causes.

The question of variation in tooth structure takes on added theoretical interest because the extreme variation (dimorphism) in tooth structure in tadpoles of *Scaphiopus bombifrons* is associated with a kind of cannibalism that is possibly adaptive. Tadpoles with one type of tooth structure and the cannibalistic behavior pattern that goes with it can survive by eating other members of the population; presumably it makes no difference to them whether the pool in which they live dries up to the point where other food supplies become insufficient. *See* Bragg (6), and Bragg and Bragg (7).

Suggested Problem and Approach

The mouthparts of tadpoles are known to change during their development after hatching. At what age or size is the definitive stage reached? Is it essentially the same for different species or markedly different? How much of this depends upon food types utilized and how much is intrinsic to each species?

Most desirable would be the collection and study of tadpoles from several different types of habitats, to supplement laboratory studies under controlled conditions. Eggs or young embryos are easier to identify than young tadpoles. Collecting eggs, growing the tadpoles in the laboratory, and comparing these with others developing in nature is one suggested procedure.

Should material from nature be unavailable (as in a large city), captive adults can be stimulated by hormones to produce their eggs, and these can be artificially inseminated. *See* Rugh (2).

The tadpoles may be reared under different conditions of temperature, food, kinds of inorganic salts in the water, and so on, and then examined at various stages for the structure of their mouthparts.

Possible Pitfalls

Positive identification of the frogs is necessary. If the animals should be of mixed species or subspecies it will be impossible to interpret the data.

Tadpoles can be preserved in 2%-5% formalin (depending upon size), but not in alcohol. (Unless formalin has been used first, alcohol will cause hardening and shrinkage.)

Larger tadpoles lose their labial denticles (teeth) as they approach metamorphosis into the adult form. This should be carefully noted, or incorrect conclusions may be drawn. If such loss is suspected, the presence or absence of ridges upon which labial denticles develop will disclose whether the denticles have been lost.

References · general

1. Rugh, Robert. 1971. A guide to vertebrate development. 6th ed. Burgess Publishing Co., Minneapolis, Minn.
2. ____. 1948. Experimental embryology. Burgess Publishing Co., Minneapolis, Minn.
3. Stebbins, R. C. 1951. Amphibians of western North America. University of California Press, Berkeley.
4. Wright, A. A. and A. H. Wright. 1947. Handbook of frogs and toads of the United States and Canada. 3d ed. Comstock Publishing Co., Ithaca, N. Y.

· specific

5. Bragg, A. N. 1961. A theory of the origin of spadefoot toads deduced principally by a study of their habits. Anim. Behavior 9:178-186.

6. ____. 1964. Further study of predation and cannibalism in spadefoot tadpoles. Herpetologica 20:17-24.

7. Bragg, A. N. and W. N. Bragg. 1958. Variations in the mouthparts of *Scaphiopus (Spea) bombifrons* Cope. Southwestern Naturalist 3:55-69.

8. Hampton, Suzanne H. and E. Peter Volpe. 1963. Development and interpopulation variability of the mouthparts of *Scaphiopus holbrooki*. Am. Midland Naturalist 70:319-328.

9. Nichols, R. J. 1937. Taxonomic studies in the mouthparts of larval Anura, Part 3. Illinois Biol. Monographs 25(4):73.

10. Smith, H. M. 1934. The amphibians of Kansas. Am. Midland Naturalist 15:377-528.

11. Wright, A. H. 1929. Synopsis of North American tadpoles. Proc. U. S. Natl. Mus. 74:1-70.

36 / LIFE HISTORY OF A SPECIES OF *PANORPA* (ORDER MECOPTERA)

Background

George W. Byers
Department of Entomology
University of Kansas
Lawrence, Kansas 66044

Nearly fifty species of scorpion-flies of the genus *Panorpa* (Mecoptera: Panorpidae) have been discovered, described, and named in North America. These principally inhabit moist woodlands in the eastern half of the United States, but some species occur as far west as Utah, with others ranging northward into southern Canada and southward into central Mexico. A few may be found in cultivated fields or pastures, especially where these adjoin woods, and some live in vegetation bordering swamps. While the immature stages of these interesting insects have been found for a few Old World species, they are virtually unknown on this continent. Larvae of one species were reported more than sixty years ago, but the complete life cycle of only one North American species has so far been made known (Byers, 1). The Mecoptera are of great interest to entomologists in that they are regarded as ancestral to all the higher groups of insects in which complete metamorphosis occurs. Accordingly, their immature stages are of considerable importance in the study of insect evolution, particularly the evolution of complete metamorphosis.

Definition of the Problem

One may obtain the larval or pupal stages of some insects by discovering where they occur in their natural environment and collecting them directly. It is necessary, however, to know that these immature forms belong to only one species, and then to rear some of them to the adult stage to determine what that species is. Alternatively, one may obtain eggs from female insects of known identity and rear the larvae (and subsequently the pupae) from these. The second method is more certain but also more difficult in most insects, yet in the case of *Panorpa* it seems the best approach, for the immature stages of these insects are diffusely scattered in the soil and have for years escaped discovery by entomologists who have sought them.

Adults of both sexes of *Panorpa* may be collected from low vegetation in the kinds of habitats just described. Males have an enlarged, bulblike structure at the tip of the abdomen and carry this curved forward over the back (hence the name scorpion-fly). In females the abdomen tapers to a slender tip. It is necessary to collect for rearing purposes males and females of the same species (and often two or three species are found together). The insects should be caged and provided with moisture, food, and resting places. A small dish of loose soil is provided for oviposition; or, if large numbers of insects are to be reared, the cage may be of a size to fit directly over a flower pot or other larger container of soil (Miyake, 4; Setty, 5). Mating and oviposition may sometimes be observed; these take place during hours of darkness in most species.

Females place their eggs beneath the surface in small clusters, usually in the upper centimeter of soil. A sample of the eggs should be recovered and preserved in 80% alcohol. When, after a few days, the eggs begin to hatch, killed insects (houseflies or *Drosophila*, easily raised in the laboratory, are adequate) or small bits of raw meat should be placed on the soil near the eggs to provide food for the young larvae. For close observation, groups of a few larvae each may be segregated into small, covered dishes provided with moist (not wet) soil and food. Damp paper tissue serves as well as soil and permits easier observation of larval activity (Byers, 1). The larvae should be given food as fast as they will consume it. Abrupt changes in size of the head capsule signify changes of larval instar; and the careful observer may, from time to time, discover a larva in the process of molting. Representatives of each of the four larval instars and of the pupal stage should be preserved in 80% alcohol.

Throughout the observations, the student should keep notes, showing the date and the particular observation made. Also, all preserved specimens should be labeled with their identity and date of preservation.

A great deal of patience, skill, and ingenuity is required to rear most insects—particularly such carnivorous insects as *Panorpa*—in the laboratory. Improvisation is necessary when experience gained from other species or genera proves inadequate. Observation of caged insects—adult or immature forms—affords an opportunity for study of their habits and discovery of many unknown aspects of their behavior.

Special Problems Involved

The life cycle of *Panorpa* in nature requires from a few months to a year, depending upon the species, but in the laboratory may be shortened to about four months by excluding the normal winter quiescence. The insects pass through the egg stage and first three larval stages rather quickly, but the last larval instar becomes inactive for a prolonged period before pupation occurs. During all this time, it is necessary to keep the soil (or paper tissue) in the rearing dishes moist, but not wet. Overheating or chilling of the larvae may be avoided by placing the rearing dishes in a cool basement. Unused food must not be left in the dishes, as it becomes overgrown with mold, and if paper tissue is used it must be periodically changed to control mold. Further suggestions concerning rearing techniques, cage construction, etc., will be found in the papers of Byers and Setty (on the closely related *Bittacus*), cited below.

References

1. Byers, G. W. 1963. The life history of *Panorpa nuptialis* (Mecoptera: Panorpidae). Ann. Entomol. Soc. Am. 56(2):142-149.
2. _____. 1970. New and little known Chinese mecoptera. J. Kans. Entomol. Soc. 43(4):383-394.
3. Carpenter, F. M. 1931. Revision of the nearctic Mecoptera. Bull. Mus. Comp. Zool., Harvard College, 72(6):205-277, plates 1-8.
4. Miyake, T. 1912. The life history of *Panorpa klugi* MacLachlan. J. Coll. Agr., Imper. Univ. Tokyo 4(2):117-139, plates 13-14.
5. Setty, L. R. 1937. Order Mecoptera, *Bittacus*. In Culture methods for invertebrate animals, pp. 335-337, Comstock Publishing Co., Ithaca, N. Y.
6. _____. 1940. Biology and morphology of some North American Bittacidae (Order Mecoptera). Am. Midland Naturalist 23:257-353.

37 / PROBLEMS IN THE REARING TECHNIQUE OF HOUSEFLIES

Background

Luther S. West
Professor of Biology, Emeritus
Northern Michigan University
Marquette, Michigan 49855

The housefly (*Musca domestica*) is much used in research, either as a test insect for evaluating insecticides, repellents, and similar preparations, or for basic research in physiology, genetics, and related fields. This has required the development of standard procedures for producing flies of known age, uniform size, and stable inheritance. Government agencies and industrial laboratories have cooperated with one another in standardizing their rearing techniques so that evaluations arrived at by one investigator can be duplicated by others, and results compared. A rather impressive literature has accumulated through the years on methods of handling fly colonies, with each new worker usually adding some feature related to the particular problem under investigation. Early investigators relied usually on natural breeding media, such as animal manures, with adult food provided in the form of bread and milk or similar "ready made" materials (2). Modern workers employ precise formulae, calculated to furnish a complete diet, properly balanced, in terms of the fly's physiological needs and assimilative powers (17).

There still remain differences of opinion on many points. A technique proved best for the rearing of a particular laboratory strain may not serve as well for another under identical environmental conditions. This has practical connotations, as when a control procedure, perfected in the laboratory, fails to be effective in the field. Such failure may not be due wholly to geographical or climatic factors. Genetic differences between wild and laboratory types can be of equal importance and often are.

The project here suggested assumes the availability of a strain with fairly uniform heredity, for which a favorable rearing technique is being sought.

Suggested Approach

Basic to further procedure is the establishment of a colony of *Musca domestica*, obtained from the region where the student lives. The population will subsequently be divided into two subcolonies, with the object of employing a major difference in technique in handling the two groups. One possible example might be the use of water-soaked cottonseed cake as a source of food for the larvae of one colony (27), while a medium based on meat and sand (21) is provided for the other. Though facilities may not permit the fixed control of temperature, relative humidity, atmospheric pressure, or intensity of light, if the two colonies are maintained side by side in the same area, environmental fluctuations will fall alike on both populations and differing results can be considered as due chiefly to controlled differences in the rearing technique.

Obtaining and Culturing the Flies

If one is near an agricultural experiment station, health department laboratory, or industrial establishment which makes insecticides, it may be possible to obtain living adults, eggs, larvae, or pupae from an established colony. If not, the best procedure is to set out one or more baited traps near a barnyard or similar situation where flies congregate. If flies are numerous, collecting with a hand net may be adequate. Should one start with collected "wild stock," rather than laboratory-reared specimens, the first step will be to get the flies to breed satisfactorily under captive conditions. One may use any convenient basic technique, such as outlined in Reference 1, Reference 11, or Reference 8.

It is important that the investigator master the technique of maintaining a normal colony of flies under conditions as sanitary as possible. Rearing cages may be constructed of simple materials, or purchased from biological supply houses such as CCM: General Biological, Inc., Chicago; Ward's Natural Science Establishment, Inc., Rochester, N. Y.; or Carolina Biological Supply Company, Burlington, N. C. Illustrations of various types of equipment may be found in Chapter XV of Reference 11 and in "Laboratory Animals and Their Housing," which appears as Chapter IV of the BSCS publication *Innovations in Equipment and Techniques for the Biology Teaching Laboratory* (Barthelemy, Dawson, and Lee, 1964).

Procedure

During the period required to produce two or three generations of your original colony, make a study of the literature and choose two techniques which differ in one or more essential ways. Using cold to immobilize the flies, count out the same number of males and females for starting each new colony. Good sized young flies (without frayed wings) should be selected, to insure a vigorous start. The original colony may now be discontinued, if space is scarce, or it can be kept in operation as a source of fresh specimens when needed. This is almost essential, if the work is to be carried on indoors, over a considerable period of time, especially in the winter time.

Record of Results

Depending on the nature of the experiment, the investigator should devise an outline or data sheet for recording such information as can subsequently be compared and interpreted. The experiment should run for not less than three full generations, preferably five or six. Under ideal conditions it is possible for a single generation to be completed in eight to ten days.

Following are examples of the type of data which may prove important:

1. Number of days from first set-up to first oviposition.
2. Number of days from first oviposition to production of first pupae.
3. Number of pupae produced. (See literature for ways of separating pupae from breeding medium.)
4. Number of days (or hours) from appearance of first pupae to emergence of first adults.
5. Sex ratio of emerged adults.
6. Average weight of males; average weight of females.
7. Percentage of pupae failing to produce flies.
8. Number of days from first emergence of flies to first laying of eggs.
9. Average number of eggs per batch.
10. Daily mortality of adult flies; of males, of females.

Other possibilities will suggest themselves. Time-taking observations, such as making of counts, should be carried out at or about the same time each day. This may usually be done in connection with the provision of food, water, and sanitary care. Besides cards or sheets designed for the planned recording of data, a notebook should be at hand in which to record any peculiar behavior or the appearance of abnormal characters, such as deviation from typical coloration, altered wing venation, and the like. Routine confirmation of expected diurnal versus nocturnal activity should also be recorded. It may be of interest to have a barometer at hand and make readings at fixed intervals for possible subsequent correlation with notes on general activity. A recording type of barometer is desirable, but periodic readings alone can have significant value. Observations of a general nature may be as frequent as the students' schedule permits. In every case the time of day should be recorded.

Conclusions

After sufficient study of the data obtained, the investigator may formulate certain conclusions and recommendations. For example, overall results may justify a statement such as: Procedure A is recommended over Procedure B for rearing larvae of the laboratory strain here studied; or The laboratory strain here studied reproduces equally well under the two (or more) procedures tested.

Pitfalls and Special Considerations

The most essential element in rearing insects is constant vigilance. Especially important is a continuous water supply, particularly where the relative humidity of the room is allowed to fluctuate. Dehydration is frequently the main cause of a high death rate among adult flies.

If the colony is started from wild stock, it is important that the population not contain a mixture of species. While it will be relatively easy to recognize blowflies of the green bottle, blue bottle, or black bottle types, care should be exercised to make sure that species more closely resembling *Musca domestica* are not included. The lesser house fly (*Fannia*), the non-biting stable fly (*Muscina*), the biting house fly (*Stomoxys*), and several species of *Sarcophaga* fall here. Also, in recent years, the face fly, *Musca autumnalis*, has appeared in most North American localities where *Musca domestica* occurs. The two species may be distinguished by consulting Sabrosky's 1959 Report.

Guard against letting flies escape into the laboratory during care of the colonies. An individual may cling to the hand or arm and thus be withdrawn from the cage. It is best to complete all care of one colony before proceeding to the next. To move back and forth between them is to invite transfer of specimens, which could obviously destroy the scientific value of the experiment. Screening must be fine enough to prevent the escape of stunted specimens. Cages should be equipped with double curtains or sleeves to reduce the chances of escape while cages are being tended. Frequent and thorough care is necessary to keep undesirable odors to a minimum.

You should practice with nonexperimental flies in perfecting your technique for immobilizing specimens without affecting their viability. This is especially important if you use an anesthetic, such as ether, rather than one of the "chilling" techniques now recommended (29 and 39).

In many cases it has been possible, in addition to the original paper, to list a source where an abstract of the content may be found. The three abstracting journals most valuable in the field of Sanitary Entomology are indicated by abbreviation as follows:

Review of Applied Entomology, Series B	RAE-B
Tropical Diseases Bulletin	TDB
Biological Abstracts	BA

References · general

1. Galtsoff, P. S., et al. 1937. Culture methods of invertebrate animals. Comstock Publishing Company, Ithaca, New York. Reprinted by Dover Press, 1959. (Houseflies, by H. H. Richardson, pp. 429-432.)
2. Hockenyos, G. L. 1931. Rearing houseflies for testing contact insecticides. J. Econ. Ent. 24(3):717-725. Abstract: RAE-B 19:197 1931.
3. Louw, B. K. 1964. Physical aspects of laboratory maintenance of muscoid fly colonies. Bull. World Health Organization 31:529-533. Abstracts: RAE-B 55:27(99) Feb., 1967. BA 47:5528 (64816) July 1, 1966; TDB 62:932, Sept., 1965.
4. Ozburn, G. W. 1964. A simplified technique for rearing and maintaining a colony of house flies (*Musca domestica* L.) Pap. Mich. Acad. Sci. 49(1):203-206. Abstract: RAE-B 54:204, Oct., 1966.
5. Sabrosky, C. W. 1959. Recognition of species of *Musca*. Cooperative Economic Insect Report 9(45): Nov., 1959. One page, illustrated. Issued by Insect Identification and Parasite Introduction Research Branch, U. S. Dept. of Agriculture, Washington, D. C.
6. Sawicki, R. M. and D. V. Holbrook. 1961. The rearing, handling and biology of house flies (*Musca domestica* L.) for assay of insecticides by the application of measured drops. Pyrethrum Post 6(2):3-18. Abstract: RAE-B 51:158, Aug., 1963.
7. Schoof, H. F. 1964. Laboratory culture of *Musca*, *Fannia*, and *Sto-*

moxys. Bull. World Health Organization 31:539-544. Abstracts: RAE-B 55:27(101) Feb., 1967; BA 47:5529 (64829), Jul., 1966; TDB: 62:932, Sept., 1965.

8. Smith, C. N. (ed.) 1967. Insect colonization and mass production. Academic Press, New York and London. (Houseflies by D. Spiller, 203-225). Abstract: RAE-B 56:1(1), Jan., 1968.

9. ———. 1967. Mass-breeding procedures. In Genetics of insect vectors of disease. J. W. Wright and R. Pal, eds. Elsevier Publ. Co., Amsterdam, Abstract: RAE-B 56:158(564), Aug., 1968.

10. Spiller, D. 1963. Procedure for rearing houseflies. Nature 199(4891): 405. Abstracts: RAE-B 52:118, Jul., 1964.

11. West, L. S. 1951. The housefly: its natural history, medical importance and control. Comstock Publ. Co., Ithaca, New York.

12. ——— and O. B. Peters. [In Press] An annotated bibliography of *Musca domestica* Linnaeus. Dawson's of Pall Mall, Cannon House, Folkestone, Kent, England.

· specific

a. *Relating to Rearing Media.*

13. Deoras, P. J. 1954. Breeding the Indian housefly (*Musca domestica nebulo* Fabr.) for experimental studies. Parasitology 44:304-309. Abstracts: RAE-B 44:14, Jan., 1956; TDB 52:221, Feb., 1955.

14. Fisher, R. W. and F. Jursie. 1958. Rearing houseflies and roaches for physiological research. Canad. Ent. 90(1):1-7. Abstract: RAE-B 47: 98, July, 1959.

15. Frings, H. 1948. Rearing houseflies and blowflies on dog biscuit. Science 107(2789):629-630.

16. Gerberick, J. B. 1948. Rearing houseflies on common bacteriological media. J. Econ. Ent. 41(1):125-126. Abstract: RAE-B 37:173, 1949.

17. Greenberg, B. 1954. A method for the sterile culture of housefly larvae, *Musca domestica* L. Canad. Ent. 86:527-528. Abstracts: RAE-B 44:113, Aug., 1956; BA 29:2978, Dec., 1955; TDB 53:1493, Dec., 1956.

18. Hafez, M. 1948. A simple method for breeding the housefly, *Musca domestica* L., in the laboratory. Bull. Ent. Res. 39(3):385-386. Abstract: RAE-B 37:57, 1949.

19. Harrison, R. A. 1949. Laboratory breeding of the housefly (*Musca domestica* L.). New Zealand J. Sci. Tech. Ser. B 30:243-247. Abstracts: BA 26:695, Mar., 1952; RAE-B 38:174, 1950.

20. Ignoffo, C. M. and I. Gard. 1970. Use of Agar-base diet and house fly larvae to assay -Exotoxin activity of *Bacillus thuringiensis*. J. Econ. Ent. 63(6):1987-1989. Abstract: RAE-B 59(6):209 (876), June, 1971.

21. Mitra, R. D. 1952. A medium for breeding of house flies, *Musca nebulo* F. in the laboratory for the study of toxicity of insecticides and resistance of flies to DDT. Sci. and Culture 17(8):341. Abstract: BA 26:2330, Sept., 1952.

22. Monroe, R. E. 1962. A method for rearing house fly larvae aseptically on a synthetic medium. Ann. Ent. Soc. Am. 55(1):140, Jan., 1962. Abstracts: RAE-B 50:243, Nov., 1962; BA 38:930 (12028), May, 1962; TDB 59:731, July, 1962.

23. Osborn, A. W. and E. Shipp. 1965. An economical method of maintaining adult diptera. J. Econ. Ent. 58(5):1023.

24. Roy, D. N. and L. B. Siddons. 1940. On continuous breeding of flies in the laboratory. Indian J. Med. Res. 28(2):621-624. Abstract: RAE-B 29:86, 1941.

25. Sawicki, R. M. 1964. Some general considerations on housefly rearing techniques. Bull. World Health Organization. 31:535-537. Abstracts: RAE-B 55:27(100), Feb., 1927; BA 47:5529(64827), Jul., 1966; TDB 62:932, Sept., 1965.

26. Tharumarajak, K. and E. S. Thevasagayam. 1961. A simple method for breeding the house-fly, *Musca domestica vicina* Macquart, in the laboratory. Bull. Ent. Res. 52(3):457-458. Abstracts: TDB 59:212, Feb., 1962; BA 39:975 (12393), 1962; RAE-B 49:282, Dec., 1961.

27. Wattal, B. L., M. L. Mannen, and N. L. Kalra. 1959. A simple medium for laboratory rearing of housefly (*Musca domestica nebula* Fabricius) with some observations on its biology. Indian J. Malariol. 13(4):175-183. Abstracts: BA 40:1935 (25387), Dec., 1962; TDB 58:141, Jan., 1961.

b. *Relating to Manipulative Techniques.*
28. Bailey, D. L., G. C. LaBrecque, and T. L. Whitfield. 1970. A forced-

air column for sex separation of adult house flies. J. Econ. Ent. 63(5): 1451-1454. Abstracts: BA 52(4):2134 (20408), Feb. 15, 1971;TDB 68:254 (473), Feb., 1971.

29. Caldwell, A. H., Jr. 1956. Dry ice as an insect anesthetic. J. Econ. Ent. 49(2):264-265. Abstract: RAE-B 45:80, May, 1957.

30. Cummings, E. C., J. T. Hallett, and J. J. Menn. 1964. A cylindrical cage for fly rearing. J. Econ. Ent. 57:177. Abstracts: BA 45:4732 (57690), Jul., 1964; TDB 61:727, Jul., 1964.

31. Dodge, H. R. 1960. An effective, economical fly trap. J. Econ. Ent. 53(6):1131-1132.

32. Hays, S. B. and G. M. Amerson. 1966. New devices for rearing and handling house flies in the laboratory. J. Econ. Ent. 59:1523-1524. Abstracts: BA 48:10177 (114361), Nov., 1967; TDB 64:437, Apr., 1967.

33. Incho, H. H. 1954. A rapid method for obtaining clean house fly pupae. J. Econ. Ent. 47(5):938-939. Abstracts: BA 29:2005, Aug., 1955; RAE-B 43:158, Oct., 1955.

34. Knipe, F. W. and H. Frings. 1952. An improved cage for flies. J. Econ. Ent. 45:1099. Abstract:RAE-B 41:109, Jul., 1953.

35. Nagasawa, S. and S. Asano. 1963. An inbreeding method of rearing the house fly. J. Econ. Ent. 56:714. Abstracts: BA 45:3219 (39869), May 1, 1964; RAE-B 52:24, Jan., 1964; TDB 61:222, Feb., 1964.

36. Schoof, H. F. 1952. The attached bait pan fly trap. J. Econ. Ent. 45: 735-736. Abstract: RAE-B 41:44, Mar., 1953.

37. Smith, A. G. 1961. Notes on breeding houseflies (*Musca domestica* L.) New Zealand J. Sci. Tech. 4(2):292-295. Abstracts: RAE-B 50:246, Nov., 1962; BA 36:8114 (86689), Dec., 1961.

38. _____ and R. A. Harrison. 1951. Notes on laboratory breeding of the housefly (*Musca domestica* L.). New Zealand J. Sci. Tech., Sec. B, 33 (1):104. Abstracts: BA 26:3222, Dec., 1952; TDB 50-161, Feb., 1953.

39. Wolf, W. W., R. A. Killough, and J. G. Hartsock. 1967. Small equipment for immobilizing flies with cool air. J. Econ. Ent. 60:303-304. Abstracts: BA 48:6036 (67470), Jul., 1967; RAE-B 107 (380), June, 1967.

38/ VARIATIONS IN THE MORPHOLOGY OF THE RESPIRATORY ORGANS OF AQUATIC INSECTS

Background

Robert G. Wetzel
W. K. Kellogg Biological Station
Michigan State University
Hickory Corners, Michigan 49060

With variations in the physical, chemical, and biological characteristics of a stream, brook, pond, or lake, marked differences occur in the qualitative and quantitative distribution of the bottom and attached fauna. Many of these benthic (bottom living) organisms have very specific requirements with respect to temperature, water chemistry, current, substrate, associated algae, other water plants, and other factors. Hence, such animals are severely restricted in the niches they are capable of occupying. Other benthic fauna possess increasingly variable tolerances to the differing aquatic conditions and are progressively more adaptable to expansion within the existing environment.

Many of the benthic organisms of the aquatic habitat consist of a large community of immature stages of insects. The tracheal respiratory system common to terrestrial insects is altered in many directions in adaptation to existence in the aquatic habitat, especially in the immature stages. Although many of the insect orders have aquatic or semiaquatic representatives, some are more conspicuous than others in this respect. Nymphs of the mayflies (*Ephemeroptera*) respire to a certain extent cutaneously (through the skin) but primarily by external articulated gills occurring laterally along the abdominal segments. Damselfly and dragonfly naiads (*Odonata*) obtain dissolved oxygen from the water mainly by means of well-developed caudal gills. Among the rocks of many streams, clinging stonefly nymphs (*Plecoptera*) may be found, many possessing external gills located variously over their bodies. The caddisfly larvae (*Trichoptera*) have many species noted for their elaborate case building and feeding mechanisms. These larvae respire cutaneously and, in some cases, with the aid of external gill appendages. The large larvae of alderflies and dobsonflies (*Megaloptera*), often prized as fish bait, have developed long lateral processes and/or extensive filaments which function as gills.

Along the shoreline of small lakes or quiet areas of streams, on the surface and among the aquatic vegetation, one can usually find numerous adult forms of water beetles (*Coleoptera*) possessing unusual respiratory adaptations to an aquatic existence. Many beetle larvae respire by means of gill appendages. Many of the aquatic larvae of the large order including the flies and mosquitoes (*Diptera*) respire by means of gills and hairlike appendages.

Suggested Problem

In general, flowing water of brooks and streams has a richer supply of dissolved oxygen than the standing water of lakes or ponds. Is there a reduction in size and in number of parts of the respiratory organs of benthic organisms, such as insect larvae that live in moving waters as compared to those inhabiting lakes? Relatively little factual data have been assembled concerning such variations intra- and interspecifically among the benthos.

Suggested Approach

In order to approach the problem in a meaningful way the investigation should include several aspects, namely: (1) the analysis of several varying aquatic habitats; (2) collection, preservation, and analysis of certain characteristics of immature aquatic insects; and (3) correlation of the factors of the environment to morphological differences found among the respiratory organs of the insects.

From the littoral areas of a lake(s) and a pond(s) make extensive collections of immature aquatic insects. The collecting devices are simple. If they are not readily available, they may be constructed with a minimum of materials (9, pp. 299-333). Specimens from rocks, debris, algal and macrophyte (water plant) stands, and similar sources other than the bottom substrates should also be carefully gathered. Exert extreme care in making concise field notes, including area sketches and a coding system correlating collected specimens with the physical and biological characteristics of the area. Preservation of the organisms is extensively discussed in the excellent volume of Usinger, *et al* (8).

Make similar collections from many differing habitats of several streams and brooks. Investigate encroaching shorelines, quiet pools, recessed areas, swifter regions and rapids with equal care.

At each collection area make quantitative determinations of several of the major relevant physical characteristics. Determine the current velocity (9, p. 151) and make a general classification of the substrate (4, pp. 242-243). Collect water samples at each site for dissolved oxygen concentration determinations, taking care to use a simple yet essential collection technique and apparatus, easily constructed from laboratory materials (9, pp. 202-207). Collected samples must be chemically fixed in the field. These can be titrated in the laboratory immediately following the field work (9, pp. 206-211).

In addition to the foregoing physical characteristics, note the types of vegetation occurring within the habitat of the insects (4, pp. 243-245), associated depths, and other attributes.

The major task in analyzing the insect forms will be that of identification, at least as to genera. Considerable patience is required, but one soon attains sufficient proficiency to key the individuals to genera and, in many cases, even to species. Such references as Chu (1) and especially Pennak (5) and Usinger (8), are all excellent for this purpose.

By very careful dissection of the external respiratory organs and by permanently mounting them on microscope slides, comparable counts, micromeasurements, and determinations of area by planimetry can be made among the same species, among species of the same genus, and among different genera in varying habitats. Attempts to correlate these data of the ratio of total surface area of the organs to body weight with physical conditions of the habitat should prove fruitful.

Special Considerations

1. Dissolved oxygen concentration of the water of the littoral areas of lakes can fluctuate considerably during periods of high illumination and darkness due to oxygen production by aquatic plants. This factor should be taken into account in the final analysis of the data (16, p. 180). The current may be strong and highly variable from one part of the stream to another and some immature insects avoid the currents among the sediments (*see*, for example, 12, 13, 15).

2. The relationship of the area of respiratory organs in some immature aquatic insects to oxygen concentrations appears to be direct (14). In some forms the function of such organs is not solely for exchange of gases but

may serve in ventilation of other sites (for example, the body surface) of absorption (17).

3. Many of the immature insects remain in the aquatic stage of their life cycle for varying periods of time. For example, the nymphs of some may-flies live in water only a few months, and others for two years, but generally most species remain for one year. Emergence usually occurs in the summer; therefore collection would be most effective in late winter, spring, and early summer for a majority of the insects.

4. The ambitious student can apply statistical procedures to the collected data to verify acceptable significance among differences found (2, 3, 7).

5. One should be cautious in arriving at any quick conclusions in relating individual physical conditions alone to morphological adaptations. An insect is an extremely complex organism. The morphological end condition of an organism at the present time is a result of the interaction of a vast array of lengthy exposures to varying ecological conditions through geological time. The organism has been constantly changing and generally becoming more specialized and specific in its habitat requirements, although in many organisms wide tolerances to extremes in environmental conditions exist.

References · general

1. Chu, H. 1949. How to know the immature insects. William C. Brown Co., Dubuque, Iowa.
2. Dixon, W. J. and F. J. Massey, Jr. 1957. Introduction to statistical analysis. McGraw-Hill Book Co., New York.
3. Lacey, O. L. 1953. Statistical methods in experimentation. The Macmillan Co., New York.
4. Lagler, K. F. 1956. Freshwater fishery biology, 2d ed. William C. Brown Co., Dubuque, Iowa.
5. Pennak, R. W. 1953. Freshwater invertebrates of the United States. The Ronald Press Co., New York.
6. Ruttner, F. 1953. The communities of running water. In Fundamentals of limnology. University of Toronto Press, Toronto, Canada.
7. Snedecor, G. W. 1956. Statistical methods applied to experiments in agriculture and biology. Iowa State College Press, Ames, Iowa.

8. Usinger, R. L., ed. 1956. Aquatic insects of California with keys to North American genera and California species. University of California Press, Berkeley.
9. Welch, P. S. 1948. Limnological methods. McGraw-Hill Book Co., New York.
10. _____. 1952. Running waters in general. In Limnology, McGraw-Hill Book Co., New York.

· specific

11. Cummins, K. W. 1962. An evaluation of some techniques for the collection and analysis of benthic samples with special emphasis on lotic waters. Am. Midland Naturalist 67:477-504.
12. Dodds, G. S. and F. L. Hisaw. 1924a. Ecological studies of aquatic insects. I. Adaptations of mayfly nymphs to swift streams. Ecology 5:137-148.
13. _____. 1924b. Ecological studies of aquatic insects. II. Size of respiratory organs in relation to environmental conditions. Ecology 5:262-271.
14. _____. 1925. Ecological studies on aquatic insects. III. Adaptations of caddisfly larvae to swift streams. Ecology 6:123-137.
15. Macan, T. T. 1963. Freshwater ecology. John Wiley & Sons, New York.
16. Wingfield, C. A. 1939. The function of the gills of mayfly nymphs from different habitats. J. Exptl. Biol. 16:363-373.

39 / THE EFFECT OF ETHYLENE GAS ON SHOOT DEVELOPMENT

Ernest Ball
Department of Botany
University of North Carolina
Raleigh, North Carolina 27607

Background

A man returning home after a stay in the country packed a carton to take back to the city. He put in two layers of fruits. The bottom layer contained some nearly ripe apples which he had just picked, and the top layer contained hard, green tomatoes. When he unpacked the carton several days later he was amazed to find that every tomato had turned beautifully red, with not a spot of green to be seen. This can be explained on the basis that the ripening apples produced the gas ethylene, an unsaturated hydrocarbon with the formula $H_2C=CH_2$, which hastened the apparent ripening of the tomatoes.

Suggested Problem

Since ethylene has an effect on the ripening of fruits, it can be classed with the plant-regulating substances or auxins. Since it affects the fruit, is it not likely that it will also affect other parts of the plant? This can lead to several problems for investigation, among which is the question: Does ethylene affect the development of the shoot (stem and leaves)?

Suggested Approach

A simple method of attack is here suggested. However, the method can be elaborated and improved, by an ingenious experimenter, into a much more sophisticated approach to the problem.

In a large bell jar, which can be sealed over a glass plate with silicone stopcock compound (or Vaseline), place a pot of moist soil containing bean or pea seeds. Any other rapidly growing dicot would serve as well. Around the pot place six or more apples that have not yet ripened fully. Set up, as a control, a comparable bell jar without the apples, and grow additional controls in air. Allow the seeds to germinate and then compare the seedlings of the experimental and control parts for such things as internode lengths, leaf area, number of stomates per unit area of leaf, and total number of stomates per leaf. Perhaps there are other comparisons that you will decide to make. Is there a difference in rate of growth? In color of leaves? In number of leaves?

Additional Problems

Perhaps this work will lead to other questions and to related problems. For example, if ethylene gas can speed ripening of fruits, are there gases that will retard ripening? In this connection, recent work at the New York State College of Agriculture has demonstrated that apples can be stored for long periods without ripening if they are kept in a controlled atmosphere composed of definite proportions of carbon dioxide, oxygen, and nitrogen. Does this atmosphere inhibit the production of ethylene? Does it destroy the ethylene as it is produced? Or is there a different mode of action involved?

References · general

1. Crocker, W., A. E. Hitchcock, and P. W. Zimmerman. 1935. Similarities in the effects of ethylene and the plant auxins. Contributions of the Boyce Thompson Institute 7(3):231-248.

2. Ferry, J. F. and H. S. Ward. 1959. Fundamentals of plant physiology. The Macmillan Co., New York.

3. Harvey, R. B. 1928. Artificial ripening of fruits and vegetables. University of Minnesota Agricultural Experiment Station Bulletin 247.

4. Isaac, W. E. 1938. The evolution of a growth-inhibiting emanation (ethylene) from ripening peaches and plums. Trans. Roy. Soc. So. Africa 26(3):307-317.

5. Kropfitsch, M. 1951. Influence of ethylene on number of stomata. Protoplasma 40:256-265.

6. Meyer, B. S., D. B. Anderson, and R. H. Böhning. 1960. Introduction to plant physiology. D. Van Nostrand Co., Princeton, N. J.

7. Taylor, H. V. 1962. Volatile compounds from fruit may affect flowers and plants. Jour. Roy. Soc. 87:187-189.

8. Thomas, M., S. L. Ranson, and J. A. Richardson. 1956. Plant Physiology. 4th ed. Philosophical Library, New York.

40/ THE LIFE HISTORY OF A PLANT

Alan A. Beetle
Plant Science Division
University of Wyoming
Laramie, Wyoming 82070

Background

Only a few plants are well known. These are mostly our important economic plants. Other plants, except for names and incomplete morphological descriptions, are poorly known. In the past, crash programs have sometimes been developed to discover the obvious facts which could easily have been recorded over the years, had there been enough observers. One example can be found in the wartime study of guayule as a possible source of rubber. Another example is the poisonous plant halogeton. No one can predict what plant may be found to have economic value or indicator significance, or exhibit unusual biological behavior.

Suggested Approach

Pick out any single plant. Any plant, even the most common, can be an open door to science. Start an FBI file. Keep a notebook of observations, measurements, and discoveries as they are made. Always keep a careful record of place and date of any observation. Develop powers of observation. Learn to recognize the plant at any season of the year, even if it may mean digging away the snow, or burrowing in the ground.

Look for answers to such questions as how and when the seeds are shed, where they go, when they germinate. What happens to the seeds that fall to make new plants? Take notes on both horizontal and vertical distribution of the dissemule.

Juvenile stages, that period following independence of the food reserve in the seed but preceding vegetative or seed reproduction, should be watched for percentage of deaths of individuals, relative development of above and below ground parts, and variations in ecotypic expression. How fast does the plant grow?

Reproductive stages should be dated for earliness, duration, and lateness over a number of seasons. When do the flowering buds appear, when do the flowers open, when does the pollen ripen, does it ripen before, while, or after the stigmas in the same flower are receptive, how is the pollen shed —if by wind in what direction, if by insects what kinds? What are the relative lengths of time for reproductive development, flowering, maturing of seed, shedding of seed? If seeds are not produced, then asexual means of reproduction will offer equally interesting points for observation. When does the plant reproduce vegetatively, when are buds produced, how are they subdivided, how reliable is the method? Take detailed notes on distribution, and when disjunct patterns occur, then learn the reason. Amass quantitative data—sizes of plants, sizes of leaves, depth of roots, and the like—but do not overlook qualitative data such as relationships with other plants, with other animals, past history in the area, and the like.

Additional Considerations

Watch for opportunities which nature is forever providing. Comparisons of development during strikingly different seasons, a wet, cold season as compared to a hot, dry season; reaction to burying; reaction to burning; reaction to grazing on two sides of a fence; changes with soil types or dominants, or elevation. Above all, take one plant species and stick with it.

Study the life histories of well-known economic plants such as wheat, tobacco, cotton, or beans. Read about them in the library and then try to learn as much or more about a wild plant.

References

1. Curtis, J. T. 1950. Outline for ecological life history study of vascular epiphytic plants. Ecology 33:550-558.
2. Humphrey, R. R. 1962. Range ecology. The Ronald Press Co., New York. pp. 1-234.
3. Pelton, J. 1951. Outline for ecological life history studies in trees, shrubs, and stem succulents. Ecology 32:334-343.
4. _____. 1953. Ecological life cycle of seed plants. Ecology 34:619-628.
5. Phillips, E. A. 1959. Methods of vegetation study. Holt, Rinehart & Winston, New York. pp. 1-107.
6. Stevens, O. A. and L. Roch. 1952. Outline for ecological life history studies of herbaceous plants. Ecology 33:415-422.

41 / THE TIME OF MOST ACTIVE CELL DIVISION IN ROOT TIPS OF PLANTS

Background

Sister Marie Bernard, O.S.F.
Biology Department
Marian College
Indianapolis, Indiana 46222
In many plants one or more periods of peak mitotic activity occur during a 24-hour period. In the onion, *Allium cepa*, these periods occur around one and eleven P.M. The factors influencing such mitotic activity are not clearly established. It would be helpful to investigate whether or not peak mitotic activity is susceptible to change by altering certain environmental factors. Such investigations may yield valuable information regarding the fundamental nature of mitotic rhythms and may have practical values in plant propagation.

Suggested Problems

1. Do various plants have the same times of peak activity? If not, what times do they have?

2. Is it possible to alter natural times of peak mitotic activity by varying such environmental factors as light and temperature?

Suggested Approach

1. Determine the time of peak mitotic activity of root tips selected from one or more species of plants kept under ordinary cycles of light and dark. This may be done by cytological examination of specimens taken at hourly intervals throughout the day and prepared as described below. The number of cells showing mitotic configurations as compared to those in interphase is a measure of relative mitotic activity.

2. Subject a selected plant to a different cycle of illumination. For example, keep it in darkness during the day and illuminate it at night.

Compare results obtained from each of the following procedures:

    a. Subject the entire plant to the altered cycle.

    b. Subject shoots to the altered cycle while maintaining roots in continuous darkness.

3. Similar experiments may be devised in which temperature cycles rather than light cycles are imposed on the plant. Further possibilities are afforded by combinations of light and temperature cycles.

Plants that produce bulbs are ideal for these studies. The bulbs can be supported over a container of water. Roots that reach a length of eight or ten millimeters are ready for cutting. Note the time at which cuttings are made. Kill and fix roots in a solution consisting of one part glacial acetic acid and three parts absolute alcohol for at least one hour.

Following are two methods for staining chromosomes.

Aceto-orcein Method:

1. Transfer root tips (with a pipette) to solution containing one part 95% alcohol and one part concentrated HCl for approximately five minutes. (Time must be watched closely; it may not take five minutes.) The root tips should be left in the alcohol acid mixture long enough to soften the tissue, but not so long as to damage it. Cells left too long will not take the stain. It is important that you determine the appropriate time for each plant type.

2. Transfer to solution made of one part glacial acetic acid, 3 parts chloroform, 6 parts absolute alcohol, for five minutes.

3. With a razor blade cut very small pieces (0.5 mm) from the tip. Place these in a small drop of aceto-orcein stain on a clean slide. Stain from 3-20 minutes. (See stain formulae at the end of this problem.)

4. Place a cover glass over the drop and press gently on the cover with the end of a matchstick. The cells should separate from each other and spread out in a single layer.

5. Seal the edges of the cover glass with melted paraffin. Examine with a microscope. A good preparation may be made permanent.

Feulgen Method:

1. After fixation, wash root tips in water and then place them in 1 normal HCl in an oven or some similar device at $60°$ C for 10-15 minutes.

2. Transfer to cool 1 N HCl.

3. Wash briefly in water and place in Reagent A "stain." (See reagent formulae at the end of this problem.) Leave until chromosomes are well

stained (10-30 minutes). Avoid exposing the stain to air any longer than necessary. The reagent must remain colorless.

4. Wash out excess Reagent A thoroughly with sulfurous acid solution (Reagent B).

5. Mount root tips in 45% acetic acid, cover, lightly heat and press out as for aceto-orcein smears.

6. Examine the preparation and if good, proceed to make permanent.

To make preparations permanent, place the slide on a block of dry ice and press it down with the eraser end of a pencil until the material is thoroughly frozen. Length of time of freezing causes no damage. As many slides as the dry-ice block will hold can be frozen at one time. Allow to freeze for at least 15 minutes.

Pry the cover slip off while the slide is still on the dry ice by slipping a razor blade or needle under a corner and lifting. The cover slip pops off easily, even if the slides have been sealed with paraffin-mastic mixtures. Nearly all the smeared material sticks to the slide when the cover slip is removed.

Place the frozen slides or cover slips immediately, before thawing, in 95% or absolute alcohol. In about five minutes they may be put in the second and last alcohol for any period from a few minutes to five hours. They are then mounted.

Formulae:

Aceto-orcein

Glacial acetic acid (65%)	30 cc
Orcein	0.3 g

(One of the best orceins is available from: Gurr and Company, London; Starkmann Laboratories, Toronto, Canada.)

Feulgen Reagent "A"

To a solution of 1 g of basic fuchsin in 200 cc of water, add 2 g of potassium metabisulfite and 10 cc of normal HCl. Allow the solution to bleach for 24 hours, add 0.5 g of nerite (activated charcoal) and shake for a minute or so, then filter rapidly through coarse filter paper. The solution should be colorless and is used as the "stain." Feulgen Reagent "A" can be purchased already prepared from several supply houses.

Feulgen Reagent "B"

Make up a sulfurous acid solution by adding 10 cc of normal HCl to a solution of 2 g of potassium metabisulfite in 200 cc of water.

References · general

1. Robertis, E. D. P. de, W. W. Nowinski, and F. Saez. 1960. General cytology. 3d ed. W. B. Saunders Co., Philadelphia.

· specific

2. Cold Spring Harbor Symposium on Quantitative Biology, Cold Spring Harbor, New York. 1960.
3. Conger and Fairchild. 1953. A quick-freeze method for making smear slides permanent. Stain Technol. 28:28.
4. DeLamater, E. D. 1948. Basic fuchsin as a nucleic stain. Stain Technol. 23:161-176.
5. Mazia, D. 1953. Cell division. Sci. Am. 53:27.
6. Prescott, D. M. 1961. The growth duplication cycle of the cell. Int. Rev. Cytol. 11:255-282.
7. Taylor, J. H. 1957. The time and mode of duplication of chromosomes. Amer. Nat. 91:209-221.
8. Vanderlyn, L. 1948. Somatic mitosis in the root tip of *Allium cepa*. Botan. Rev. 14:270-318.
9. Van't Hof, J. 1965. Relationships between mitotic cycle duration, S-period duration and the average rate of DNA synthesis in the root meristem cells of several plants. Exp. Cell Res. 39:48-58.

42 / GROWTH AND SYNTHESIS OF METABOLITES BY ISOLATED LEAF TISSUE

Richard H. Nieman
U. S. Department of Agriculture
P. O. Box 672
Riverside, California 92502

Background

Underlying the development of a mature organism from a fertilized egg is an orderly sequential development of numerous metabolic processes. As the organism or organ develops, new enzymes are synthesized or become activated. Studies that reveal the sequence of enzyme synthesis and the factors controlling this sequence are useful in that they help us to understand various aspects of development. One method for the study of development in plants involves the use of discs of isolated leaf tissue.

When small isolated segments of young leaves are incubated on a simple, defined medium in a moist environment, the segments continue to grow and carry on at least some of the metabolic functions of the intact leaf for a period of 48 hours or more. Discs about 5 millimeters in diameter, punched from leaves with a cork borer, are convenient tissue segments with which to work. Growth can be measured both as an increase in the diameter of the discs and as an increase in their fresh weight. This simple system provides a convenient means of studying a number of aspects of growth and metabolism of leaves under closely controlled conditions. The leaf discs can be "fed" compounds of metabolic interest and the growth rate measured.

Growth of the discs is largely the result of an increase in the size of cells in the interveinal mesophyll tissue. Cell division may occur if the discs are given light, sugar, and nitrate. The veins show little or no growth. Some factors that will stimulate growth of the discs are light, adenine, yeast, nucleic acid, gibberellin, kinetin, aqueous extracts of leaves and seeds, and the cobaltous ion. Indoleacetic acid does not appear to affect mesophyll tissue, but does stimulate growth of the veins. At least part of the stimulating effect of extracts from leaves and seeds appears to be attributable to their adenine content.

While the synthesizing properties of the discs have not been examined in any detail, some preliminary investigations indicate that the ability to synthesize protein and nucleic acids develops as the leaf emerges from its bud and begins to enlarge. The ability to synthesize other compounds probably follows a similar pattern. This could be investigated by analyzing the metabolic responses and synthesizing powers of discs taken from leaves of various ages.

Suggested Problems

A study employing leaf punches might be directed along the following lines.

1. The response of the discs to a "diet" of selected inorganic ions and organic compounds.

2. The synthesis of metabolites by discs taken from leaves of different ages and incubated on different media.

Suggested Approaches and Procedures

The first problem can be attacked by incubating the discs for 30 to 48 hours on a base medium, for example, 1% sucrose in 0.06M KNO_3, to which the material to be tested has been added. Growth can be measured either as an increase in the diameter of the discs or as an increase in their fresh weight. A comparison of the growth of the discs on the base medium with that on base medium plus the material to be tested provides a measure of the stimulating or inhibiting properties of the added material.

The second approach—that is, an examination of the synthesis of metabolics by the discs during incubation—would be guided by the investigator's background and by the equipment available for the isolation and determination of compounds of interest. If a colorimeter is available, the protein can be determined by means of the biuret reaction. Protein can be extracted from the discs by using a solution of 0.3 to 1.0 N NaOH or KOH after a preliminary extraction with organic solvents and/or boiling water to remove interfering substances.

The flavonoids would also be an interesting group of compounds with which to work. The function of the flavonoids is not definitely known. Some of these compounds appear to have growth-stimulating properties. Many of them can be extracted directly from fresh leaf tissue with boiling water. They lend themselves very well to separation and identification by means of paper chromatography. For this, simple solvents and tall graduates or similar vessels such as chromatography jars may be used.

The equipment needed for a leaf-disc study is minimal, especially if the study is confined to growth measurements. In this case, a few petri dishes and other common laboratory glassware are adequate. The analysis of compounds synthesized by the discs can also be carried out with fairly simple equipment. A colorimeter or spectrophotometer is desirable, but not necessary. The presence or formation of flavonoids and a number of other pigments may be assessed qualitatively by means of paper chromatography.

The primary leaves of the bean plants, *Phaseolus vulgaris* L., are commonly employed as a source of leaf material. The plants grow readily in flats of sand or vermiculite, and the primary leaves are ready for use in about eight days. A comparison of the growth of discs from leaves of several plant species would be interesting.

Possible Pitfalls

Every effort should be made to maintain aseptic conditions during the incubation of the discs, especially when the synthesis of compounds is to be determined. Unless the discs are incubated in laboratories where air-borne spores are troublesome (as, for example, in laboratories where micro-organisms are cultured), aseptic conditions are not difficult to maintain for the 30- to 48-hour periods of incubation. The media can be sterilized by heating in an autoclave or pressure cooker, or by filtration. The petri dishes used as incubation vessels can be sterilized by heat. Surface disinfection of the discs is usually accomplished by gentle washing with a mild detergent (1% aqueous solution) and a rinsing with sterile water.

References · general

1. Bonner, J. F. and A. W. Galston. 1952. Principles of plant physiology. W. H. Freeman & Co., San Francisco.
2. Galston, A. W. 1961. The life of the green plant. Prentice-Hall, Inc., Englewood Cliffs, N. J.
3. Milthorpe, F. L., ed. 1956. The growth of leaves. Academic Press, New York.

4. Peach, K. and M. V. Tracey, eds. 1955, 1956. Modern methods of plant analysis. 4 vols. Springer-Verlag, Berlin.
5. Ruhland, W., ed. 1956. Encyclopedia of plant physiology. Vol. 2. General physiology of the plant cell. Springer-Verlag, Berlin.
6. White, P. R. 1954. The cultivation of animal and plant cells. The Ronald Press Co., New York.

· specific

7. Bonner, D. M., A. J. Haagen-Smit, and F. W. Went. 1939. Leaf growth hormones. I. A bio-assay and source for leaf growth factors. Botan. Gaz. 101:128-144.
8. Geissman, T. A. 1955. Anthocyanins, chalcones, aurones, flavones, and related water-soluble plant pigments. In K. Peach and M. V. Tracey, eds., Modern methods of plant analysis, 3:450-498. Springer-Verlag, Berlin.
9. Humphries, E. C. and A. W. Wheeler. 1960. The effects of kinetin, gibberellic acid, and light on expansion and cell division in leaf discs of dwarf bean (*Phaseolus vulgaris*). J. Exptl. Bot. 11:81-85.
10. Miller, C. O. 1956. Similarity of some kinetin and red light effects. Plant Physiol. 31:318-319.
11. Nieman, R. H. and L. L. Poulsen. 1967. Growth and synthesis of nucleic acid and protein by excised radish calyledons. Plant Physiology 42:946-952.
12. Powell, R. D. and Mildred M. Griffith. 1960. Some anatomical effects of kinetin and red light on discs of bean leaves. Plant Physiol. 35:273-275.
13. Scott, R. A., Jr. and J. L. Liverman. 1956. Promotion of leaf expansion by kinetin and benzylaminopurine. Plant Physiol. 31:321-322.
14. Stephenson, Mary L., K. V. Thimann, and P. C. Zamecnik. 1956. Incorporation of C^{14} amino acids into proteins of leaf discs and cell-free fractions of tobacco leaves. Arch. Biochem. Biophys. 65:194-209.

43 / THE INFLUENCE OF ORIENTATION ON THE GROWTH RATE OF TREE STEMS

Background

Arthur H. Westing
Department of Biology
Windham College
Putney, Vermont 05346

It is well known that orientation with respect to gravity influences the growth rate of tree stems and branches. Thus, if a stem is bent (trained) downward from its normal position its rate of growth will be markedly inhibited. Often such reorientation, interestingly enough, also results in more abundant flowering on the treated stem and thus greater fruit set; this knowledge has long been put to practical use in the culture of vanilla beans and of various fruit trees. Furthermore, when the terminal shoot (leader) of a tree is tied over it can no longer exert its normal apical dominance over the branches below. Under certain conditions shoot reorientation, curiously, has even resulted in a hastened breaking of winter dormancy. The phenomena described above are grouped under the heading of *geotonus*, or sometimes geomorphism or gravimorphism. (Geotonus is distinguished from the closely related phenomenon of geotropism in not being a directed or oriented growth response.)

Several recent articles—Longman and Wareing (4), Wareing and Nasr (9, 10), VanHaverbeke and Barber (8)—present the results of interesting investigations of one aspect or another of geotonus. They are useful in providing us with some quantitative data concerning the magnitudes of growth inhibition to be expected for several species of trees treated in various ways.

The shoot growth inhibition and other changes imposed by disorientation suggest an interference with the normal internal hormonal mechanisms and relationships that exist within a tree. Perhaps there is an interference with the synthesis or with the downward translocation of a plant growth hormone (auxin) such as indoleacetic acid; this auxin is presumed to be manufactured in the tip region (apical meristem) or leaves. Alternatively the possibility exists that certain factors arising in the roots are prevented from being transported upward. This suggestion stems from the indication that only a small lower portion of the stem need be disoriented for the entire portion above it to have its growth depressed. *See* McLean (5) and Wareing and Nasr (9).

Suggested Studies

It would be of great interest to determine the relation between the degree (i.e., angular extent) of downward bending of a shoot and the amount of its growth depression. (Will the depression be a function of the cosine of the angle of deviation from the vertical?) A suitable experiment could well be performed on the leaders of a group of potted one-year-old tree seedlings or on the branches within one whorl of perhaps a pine sapling growing in the open. Shoots being trained must be firmly but gently tied to supports constructed for the purpose. Precise weekly measurements of both longitudinal growth and diameter growth (using a vernier caliper) would be necessary. Untreated shoots must, of course, also be measured for comparative purposes. (Will the overall geotonic growth depression at the end of a growing season be the result of decreases in rate or duration? Will longitudinal growth depression be the result of shorter or fewer nodes?)

A logical extension of these experiments is to train stems or branches in various zigzag patterns so as to disorient only certain portions of them. In some instances the apical region alone could be reoriented with respect to gravity, in others the basal region, and in still others only a central region. To what extent does the geotonic growth depression of a disoriented region influence the growth of unchanged adjacent regions above and below?

A different approach to the problem would be to attempt to restore the normal growth rates of geotonically inhibited shoots by supplying them with growth hormones, since these can be postulated to have been reduced by the treatment. Indoleacetic acid, gibberellic acid, and kinetin are three commercially available compounds that should be used since they have their counterparts in the plant. The chemicals could be smeared on in a lanolin carrier, perhaps in a concentration of ½% by weight. Applications could be made above, at, or below a region of disorientation, singly and in various combinations.

It must be noted that when shoots particularly of deciduous trees (hardwoods) are bent down, buds are likely to grow out on the upper side. It may be necessary to pinch off these buds in order to avoid a changed overall growth pattern that would lead to a confounding of the results. Certain other problems of technique that might be encountered are dealt with in the studies that have been cited. An excellent source of more general background information is *Green Plant* (3).

Conclusion

The problem, in short, is one of adding to the scant knowledge and meager quantitative data presently available on the deceptively simple phenomenon of geotonus. That a change in orientation with respect to gravity can retard plant growth has long been known. But the subject has been little studied and the opportunities for research are numerous. The physiological basis of this fundamental plant growth phenomenon is a mystery and a challenge.

References

1. Dunn, Stuart. 1970. Light quality effects on the life cycle of common purslane. Weed Sci. 18(5):611-613.
2. Fosket, Elizabeth Baker and W. R. Briggs. 1970. Photosensitive seed germination in Catalpa Speciosa. Bot. Gaz. 131(2):167-172.
3. Galston, A. W. 1968. Green plant. Prentice-Hall, Inc., Englewood Cliffs, New Jersey.
4. Longman, K. A. and P. F. Wareing. 1958. Gravimorphism in trees: effect of gravity on flowering and shoot growth in Japanese larch (*Larix leptolepis*, Murray). Nature (London) 182:380-381.
5. McLean, F. T. 1939-1941. Loop method of dwarfing plants and inducing flowering. Contrib. Boyce Thompson Inst. (Yonkers, N.Y.) 11:123-125.
6. Ortega, J. K. E. and R. I. Gamow. 1970. Phycomyces: habituation of the light growth response. Science 168(3937):1374-1375.
7. Pearson, J. A. and P. F. Wareing. 1970. Polysomal changes in developing wheat leaves. Planta (Berlin) 93(4):309-313.
8. VanHaverbeke, D. F. and J. C. Barber. 1961. Less growth and no increased flowering from changing slash pine branch angle. Southeastern Forest Experiment Station, Asheville, N. C., Research Note. No. 167.
9. Wareing, P. F. and T. A. A. Nasr. 1958. Gravimorphism in trees: effects of gravity on growth, apical dominance and flowering in fruit trees. Nature (London) 182:379-380.
10. _____. 1961. Gravimorphism in trees. I. Effects of gravity on growth and apical dominance in fruit trees. Ann. Botany (London) 25:321-340.

APPENDIX A

CONTENTS — SERIES 2

BEHAVIOR

ANIMALS

How do the reactions of insects to the wavelength and intensity of light vary by species or stage of development, and can they be altered?

A plan is outlined for the study of factors influencing caste differentiation in termites, using controlled experiments on small laboratory groups of these social insects.

Why do aphids colonize only certain parts of their host plants?

Can you design a chemical "flyswatter" that capitalizes on the fly's own behavior?

Members of ant colonies communicate with each other by chemical substances known as pheromones. This system of communication is a fitting subject for imaginative original research.

Characteristics of innate behavior may be analyzed by observing web-building spiders.

PLANTS

PHYSIOLOGY

ANIMALS

PLANTS

MICROORGANISMS

ECOLOGY

ANIMALS

MICROORGANISMS

MICROBE/PLANT INTERACTION

MICROBIAL PHYSIOLOGY

GENETICS

PLANTS

CONTENTS — SERIES 3

BEHAVIOR

ANIMALS

175

MICROBIOLOGY

ANIMALS

GENETICS

ANIMALS

APPENDIX B — BIBLIOGRAPHY

A student who begins to look up the literature provided by the authors of the research problems may well find himself a victim of the information explosion: there are so many professional scientific journals published today that the library of even large state universities cannot have them all.

Every library, however, is part of an interlibrary loan system and has the capability to acquire a copy of an article if you can supply correct, and verifiable, bibliographic information to identify it precisely.

The following list of books and periodicals is intended to introduce you to the literature available in your chosen field of research. The list is not complete. It is merely a sampler, an indication of the rich and varied literature in a few of the many fields of biology.

If you cannot locate the books you desire, you may be able to locate others just as useful. Most textbooks contain valuable lists of references and suggestions for further reading. You might turn to these lists if you cannot find the titles that you need in the following pages. A published piece of scientific research usually includes a bibliography of literature relating more or less directly to the research in question.

Biological Abstracts is an especially useful journal that will aid you in locating references on specific subjects. Each issue contains brief summaries, arranged according to subjects of research articles published in a wide variety of periodicals throughout the world. The subject index of *Biological Abstracts* will help you to check rapidly on what has been published in your area of special interest.

Technical information generated from government research is indexed and made available in a number of ways. The easiest way to secure this information is to contact the reference librarian or government documents librarian at nearby public libraries and university libraries.

Reading these references will help you broaden, as well as intensify, your understanding of the scientific problem being researched. In turn, each reference that you read will have its own bibliography from which you may choose to read articles that may further your understanding. Review articles are particularly helpful. They usually contain comprehensive bibliographies covering a specific field of research.

There is still another source of help. The authors of research articles usually have reprints of their papers, which they distribute on request. The name and address of each author is published along with his paper. You may write to the author and request a reprint if the journal containing the article is not readily available.

GENERAL

A. Periodicals

American Naturalist
American Scientist
Audubon Magazine
Biological Abstracts
Biological Bulletin
Biological Reviews
Biologist
Bioscience
Brookhaven Symposia in Biology
Cold Spring Harbor Symposia in Quantitative Biology
Endeavour
Frontiers
Habitat
Journal of Experimental Biology
Lancet
Marine Biology
Molecular Biology
National Geographic
Natural History
Proceedings of the National Academy of Science
Proceedings of the Society for Experimental Biology and Medicine
Quarterly Journal of Microscopical Science
Quarterly Review of Biology
Science
Scientific American

B. General References

Abercrombie, M. M., *et al.* 1962. A dictionary of biology. Aldine Pub. Co., Chicago.
Arena, V. 1971. Ionizing radiation and life: an introduction to radiation biology and biological radiotracer methods. C. V. Mosby Co., St. Louis.

A short history of science: origins and results of the scientific revolution. 1959. Doubleday Anchor Books, New York.

Asimov, I. 1968. Intelligent man's guide to the biological sciences. Washington Square Press, New York.

Baer, A. S., *et al.* 1971. Central concepts of biology. The Macmillan Co., New York.

Bates, M. 1960. Forest and the sea. Random House, New York.

———. 1963. Animal worlds. Random House, New York.

Becker, G. C. 1972. Introductory concepts in biology. The Macmillan Co., New York.

Biology data book. 1964. Federation of American Societies for Experimental Biology, Washington, D. C.

Brownowski, J. 1953. The common sense of science. Harvard University Press, Cambridge, Mass.

Butterfield, H. 1965. The origins of modern science (1300-1800). Free Press, New York.

Chauvin, R., *et al.* 1973. Encyclopedia of the life sciences. 8 vols. Doubleday, New York.

Dampier, W. C. D. 1949-1966. History of science. Cambridge University Press, New York.

Gabriel, M. L. and S. Fogel, eds. 1955. Great experiments in biology. Prentice-Hall, Inc., Englewood Cliffs, New Jersey.

Glasston, S. 1967. Sourcebook on atomic energy. 3d ed. D. Van Nostrand Co., Princeton, New Jersey.

Gray, P. 1970. The encyclopedia of the biological sciences. 2d ed. Van Nostrand Reinhold Co., New York.

———. 1967. A dictionary of the biological sciences. Van Nostrand Reinhold Co., New York.

Hardin, G. J. 1966. Biology: its principles and implications. W. H. Freeman and Co. Publishers, San Francisco.

Jacker, C. 1971. The biological revolution: a background book on making a new world. Parent's Magazine, New York.

Kavlaer, L. Freezing point: cold as a meter of life and death. John Day, New York.

Marx, W. 1967. The fragile ocean. Ballantine Books, New York.

McGraw-Hill encyclopedia of science and technology. 1960. McGraw-Hill Book Co., New York.

Nordenskiöld, A. E. 1960. The history of biology. Tudor Pub. Co., New York.

Osborn, F. 1968. Our plundered planet. Pyramid Books, New York.

Painter, J. H. 1972. Biology today. CRM Books, Del Mar, California.

Phillips, E. A. 1971. Basic ideas in biology. The Macmillan Co., New York.

Simpson, G. G. 1971. Biology and man. Harcourt Brace Jovanovich, New York.

Stein, B. 1971. Dictionary of biology. Barnes & Noble, New York.

VanNorman R. W. 1970. Experimental biology. 2d ed. Prentice-Hall, Inc., Englewood Cliffs, New Jersey.

Weisz, P. B. 1971. Science of biology. McGraw-Hill Book Co., New York.

ANIMAL BEHAVIOR

A. Periodicals

American Naturalist

Animal Behaviour

Behaviour

Behavioral Science

Journal of Applied Psychology

Journal of Comparative Physiology and Psychology

Journal of Experimental Psychology

B. General References

Altman, S. A. 1967. Social communication among primates. University of Chicago Press, Chicago.

Carthy, J. D. 1971. Introduction to the behavior of vertebrates. Hafner Pub. Co., New York.

Davis, D. E. 1966. Integral animal behavior. The Macmillan Co., New York.

Dethier, V. G. and E. Stellar. 1970. Animal behavior. 3d ed. Prentice-Hall, Inc., Englewood Cliffs, New Jersey.

Droscher, V. B. 1970. The friendly beast: latest discoveries in animal behavior. E. P. Dutton & Co., New York.

Eibl-Eibesfeldt, I. 1970. Ethology: the biology of behavior. Holt, Rinehart & Winston, New York.

Etkin, W. 1964. Social behavior and organization among vertebrates. University of Chicago Press, Chicago.

_____ and D. G. Freedman. 1967. Social behavior from fish to man. University of Chicago Press, Chicago.

Evans, S. M. 1968. Studies in invertebrate behavior. Hillary House Publishers, New York.

Fink, H. K. 1972. Mind and performance: a comparative study of learning in mammals, birds and reptiles. Greenwood Pub. Corp., Westport, Connecticut.

Klopfer, P. H. and J. P. Hailman. 1967. An introduction to animal behavior: ethology's first century. Prentice-Hall, Inc., Englewood Cliffs, New Jersey.

Lorenz, K. 1952. King Solomon's ring. Apollo Editions, Inc., New York.

_____. 1966. On aggression. Harcourt Brace Jovanovich, New York.

_____. 1966. Evolution and modification of behaviour. Methuen, London.

_____. 1971. Studies in animal and human behaviour. Vol. 2. Harvard University Press, Cambridge, Massachusetts.

Marler, P. R. and W. J. Hamilton, Jr. 1966. Mechanisms of animal behavior. John Wiley & Sons, Inc., New York.

Ruwet, J. C. 1972. Introduction to ethology: the biology of behavior. International Universities Press, New York.

Schiller, C. H., ed. 1964. Instinctive behavior: the development of a modern concept. International Universities Press, New York.

Scott, J. P. 1972. Animal behavior. 2d ed. University of Chicago Press, Chicago.

Thorpe, W. H. 1963. Learning and instinct in animals. Harvard University Press, Cambridge, Massachusetts.

Tinbergen, N. 1965. Social behavior in animals: with special reference to vertebrates. 2d ed. John Wiley & Sons, Inc., New York.

_____. 1969. Study of instinct. Oxford University Press, New York.

Van der Kloot, W. G. 1968. Behavior. Holt, Rinehart & Winston, New York.

Van Lawick-Goodall, J. 1971. In the shadow of man. Houghton Mifflin Co., Boston.

Wheeler, W. M. 1969. Social life among the insects: being a series of lectures delivered at the Lowell Institute in Boston in March, 1922. 1969 Reprint of 1923 ed. Johnson Repr., New York.

ANIMALS

A. Periodicals

American Zoologist
Animal Behaviour
Animal Kingdom
Animal Life
Animals
Behaviour
Endocrinology
Fauna
Herpetologica
Herpetological Review
Journal of Biological Sciences
Journal of Experimental Biology
Journal of Experimental Marine Biology and Ecology
Journal of Experimental Zoology
Journal of Fish Biology
Journal of Mammology
Journal of Marine Research
Journal of Molecular Biology
Journal of Morphology
Journal of Protozoology
Journal of Theoretical Biology
Journal of Zoology
Systematic Zoology
Zoology Reviews
Zoologica

B. General References

Anderson, H. T., ed. 1969. The biology of marine mammals. Academic Press, New York.

Bonner, J. T. 1955. Cells and societies. Princeton University Press, Princeton, New Jersey.

Cheng, T. C. 1964. The biology of animal parasites. W. B. Saunders Company, Philadelphia.

Edmonson, W. T., ed. 1959. Ward and Whipple's freshwater biology. 2d ed. John Wiley & Sons, Inc., New York.

Grzimek, B., ed. 1972-1974. Grzimek's animal life encyclopedia. Van Nostrand Reinhold Co., New York.

Hegner, R. W. 1967. Parade of the animal kingdom. The Macmillan Co., New York.

_____ and K. A. Stiles. 1969. College zoology. 8th ed. The Macmillan Co., New York.

Hickman, C. P. 1970. Integrated principles of zoology. 4th ed. C. V. Mosby Co., St. Louis.

Moore, J. A. 1957. Principles of zoology. Oxford University Press, New York.

Stonehouse, B. 1971. Animals of the Arctic: the ecology of the far north. Holt, Rinehart & Winston, New York.

Storer, T. I. and R. L. Usinger. 1965. General zoology. 4th ed. McGraw-Hill Book Co., New York.

_____. 1968. Elements of zoology. 3d ed. McGraw-Hill Book Co., New York.

Strehler, B. L. 1970. Time, cells, and aging. Academic Press, New York.

Turner, C. D. 1971. General endocrinology. 5th ed. W. B. Saunders Company, Philadelphia.

C. Invertebrates

Barnes, R. D. 1968. Invertebrate zoology. 2d ed. W. B. Saunders Company, Philadelphia.

Barrington, E. J. 1968. Invertebrate structure and function. Houghton Mifflin Company, New York.

Bayer, F. M. and H. B. Owre. 1968. The free-living lower invertebrates. The Macmillan Co., New York.

Beklemishev, V. N. 1970. Principles of comparative anatomy of invertebrates. 2 vols. University of Chicago Press, Chicago.

Bird, A. F. 1971. The structure of nematodes. Academic Press, New York.

Brown, F. A. 1967. Selected invertebrate types. John Wiley & Sons, Inc., New York.

Bullock, T. H. and G. A. Horridge. 1965. Structure and function in the nervous systems of invertebrates. 2 vols. W. H. Freeman and Co., Publishers, San Francisco.

Buschbaum, R. M. 1948. Animals without backbones: an introduction to the invertebrates. University of Chicago Press, Chicago.

Chandler, A. C. and C. P. Read. 1961. Introduction to parasitology. 10th ed. John Wiley & Sons, Inc., New York.

Clark, A. M. 1962. Starfishes and their relations. British Museum, London.

Cloudsley-Thompson, J. L. 1958. Spiders, scorpions, centipedes and mites. Pergamon Press, New York.

Croll, N. A. 1970. The behaviour of nematodes: their activities, senses and responses. St. Martin's Press, Inc., New York.

Dales, R. P. 1963. The annelids. Hutchinson University Library, London.

Giese, A. C. 1973. The biology of light sensitive protozoa. Stanford University Press, Stanford, Conn.

Goodey, T. 1951. Soil and freshwater nematodes. John Wiley & Sons, Inc., New York.

Hall, R. P. 1953. Protozoology. Prentice-Hall, Inc., Englewood Cliffs, New Jersey.

Hegner, R. W. and J. G. Engemann. 1968. Invertebrate zoology. The Macmillan Co., New York.

Hickman, C. P. 1973. Biology of invertebrates. 2d ed. C. V. Mosby Co., St. Louis.

Hyman, L. H. 1951-1967. The invertebrates. 6 vols. McGraw-Hill Book Co., New York.

Jenkins, M. M. 1970. Animals without parents. Holiday House, New York.

Kalstner, A. 1967-1968. Invertebrate zoology. 3 vols. John Wiley & Sons, Inc., New York.

Kudo, R. R. 1954. Protozoology. 4th ed. Charles C. Thomas, Springfield, Illinois.

Lenhoff, H. M., ed. 1971. Experimental coelenterate biology. University of Hawaii Press, Hawaii.

―― and W. F. Loomis. 1961. Biology of hydra and some other coelenterates. University of Miami Press, Coral Gables, Florida.

Noble, E. R. and G. A. Noble. 1964. Parasitology. 2d ed. Lea & Febiger, Philadelphia.

Russel-Hunter, W. D. 1969. Biology of higher invertebrates. The Macmillan Co., New York.

――. 1968. Biology of lower invertebrates. The Macmillan Co., New York.

Smith, J. E., ed. 1971. The invertebrate panorama. Universe Books, New York.

Snow, K. R. 1970. The arachnida: an introduction. Columbia University Press, New York.

Zuckerman, B. M., et al. 1971. Plant parasitic nematodes. 2 vols. Academic Press, New York.

D. Vertebrates

Blair, W. F. 1968. Vertebrates of the United States. 2d ed. McGraw-Hill Book Co., New York.

Boorer, M. 1971. Mammals of the world. Grosset & Dunlap, New York.

Bularis, A. 1970. The world of reptiles. Universal Books, New York.

Farner, D. S. and J. R. King, eds. 1971. Avian biology. Academic Press, New York.

Goin, C. J. and O. B. Goin. 1971. Introduction to herpetology. 2d ed. W. H. Freeman and Co. Publishers, San Francisco.

Hoar, W. S. and D. J. Randall, eds. 1971. Fish physiology, Vol. VI. Environmental relations and behavior. Academic Press, New York.

Mochi, U. and T. D. Carter. 1971. Hoofed mammals of the world. Charles Scribner's Sons, New York.

Orr, R. T. 1971. Vertebrate biology. W. B. Saunders Company, Philadelphia.

Romer, A. S. 1971. The vertebrate body. 3d ed. W. B. Saunders Company, Philadelphia.

Schultz, A. H. 1969. The life of primates. Universe Books, New York.

Torrey, T. W. 1971. Morphogenesis of the vertebrates. 3d ed. John Wiley & Sons, Inc., New York.

Yapp, W. B. 1970. The life and organization of birds. American Elsevier Pub. Co., New York.

BIOCHEMISTRY

A. Periodicals

Advances in Enzymology
Annual Review of Biochemistry

Archives of Biochemistry and Biophysics
Biochemical Journal
Biochemistry
Journal of Biological Chemistry
Journal of Biophysical and Biochemical Cytology
Journal of Molecular Biology

B. General References

Ashby, J. F., *et al.* 1971. Principles of biological chemistry. F. A. Davis Co., Philadelphia.

Baker, J. J. and G. E. Allen. 1970. Matter, energy and life: an introduction for biology students. 2d ed. Addison-Wesley Pub. Co., Reading, Mass.

Baldwin, E. 1962. Nature of biochemistry. 2d ed. Cambridge University Press, New York.

Baserga, R. 1969. Biochemistry of cell division. Charles C. Thomas, Springfield, Illinois.

Bell, G. H., *et al.* 1969. Textbook of physiology and biochemistry. 7th ed. Williams & Wilkins Co., Baltimore.

Berger, M. 1971. Enzymes in action. T. Y. Crowell Co., New York.

Bonner, D. M. 1961. Control mechanisms in cellular processes. Ronald Press, New York.

Bonner, J. and J. E. Varner, eds. 1965. Plant biochemistry. Academic Press, New York.

Bresler, S. E. 1970. Introduction to molecular biology. Academic Press, New York.

Bull, H. B. 1971. An introduction to physical biochemistry. 2d ed. F. A. Davis Co., Philadelphia.

Dagley, S. and D. E. Nicholson. 1970. An introduction to metabolic pathways. John Wiley & Sons, Inc., New York.

Fargo, P. and J. Lagnado. 1972. Life in action: biochemistry explained. A. A. Knopf, Inc., New York.

Fearon, W. R. 1961. Introduction to biochemistry. 4th ed. Academic Press, New York.

Frieden, E. and H. Lipner. 1971. Biochemical endocrinology of the vertebrates. Prentice-Hall, Inc., Englewood Cliffs, New Jersey.

Mahler, H. R. and E. H. Cordes. 1971. Biological chemistry. 2d ed. Harper & Row, New York.

Mandlestam, J. and K. McQuillen, eds. 1968. Biochemistry of bacterial growth. Blackwell Scientific Publishers, Oxford, England.

Parpart, A. K. 1971. Chemistry and physiology of growth. Kennikat Press, Inc., Port Washington, New York.

Pauling, L. C. and H. A. Ithano, eds. 1961. Molecular structure and biological specificity. 3d ed. American Institute of Biological Sciences, Washington, D. C.

Pauling, L. C. and R. Hayward. 1970. Architecture of molecules. W. H. Freeman and Co. Publishers, San Francisco.

Thach, R. E. and M. R. Neuburger. 1971. Research techniques in biochemistry and molecular biology. W. A. Benjamin Co., Menlo Park, California.

West, E. S., *et al.* 1966. Textbook of biochemistry. 4th ed. The Macmillan Co., New York.

White, A., *et al.* 1968. Principles of biochemistry. 4th ed. McGraw-Hill Book Co., New York.

Yudkin, M. and R. Offord. 1971. Harrison's guidebook to biochemistry. Cambridge University Press, New York.

CELL BIOLOGY

A. Periodicals

Cell and Tissue Kinetics
Cytologia
Experimental Cell Research
International Review of Cytology
Journal of Cell Biology
Journal of Cellular and Comparative Physiology
Stain Technology
Transactions of the American Microscopical Society

B. General References

Albertsson, P. A. 1970. Partition of cell particles and macromolecules. 2d ed. John Wiley & Sons, Inc., New York.

Allen, J. M., ed. 1962. The molecular control of cellular activity. McGraw-Hill Book Co., New York.

———. 1970. Molecular organization and biological function. Harper & Row, New York.

Ambrose, E. J. and D. M. Easty. 1970. Cell biology. Addison-Wesley Pub. Co., Reading, Mass.

Bloom, W. and D. W. Fawcett. 1968. A textbook of histology. 9th ed. W. B. Saunders Company, Philadelphia.

Bonner, J. T. 1966. Cells and societies. Antheneum, New York.

Brachet, J. and A. E. Mirsky, eds. 1964. The cell: biochemistry, physiology, morphology. 6 vols. Academic Press, New York.

Dowben, R. M. 1971. Cell biology. Harper & Row, New York.

———. 1969. Biological membranes. Little, Brown & Co., Boston.

du Praw, E. J. 1968. Cell and molecular biology. Academic Press, New York.

Fawcett, D. W. 1966. An atlas of fine structure. The cell. W. B. Saunders Company, Philadelphia.

From cell to organism. Readings from Scientific American. 1967. W. H. Freeman and Co., Publishers, San Francisco.

Giese, A. C. 1968. Cell physiology. 3d ed. W. B. Saunders Company, Philadelphia.

Ham, A. W. and T. S. Leeson. 1969. Histology. 6th ed. J. B. Lippincott Co., Philadelphia.

Harris, P. J., ed. 1971. Biological ultrastructure: the origin of cell organelles. Oregon State University Press, Corvallis, Oregon.

Jensen, W. A. 1970. The plant cell. 2d ed. Wadsworth Pub. Co., Belmont, California.

Ledbetter, M. C. and K. R. Porter. 1970. Introduction to fine structure of plant cells, Springer-Verlag, New York.

Lowey, A. G. and P. Siekevitz. 1966. Cell structure and function. Holt, Rinehart & Winston, New York.

McElroy, W. D. 1971. Cell physiology and biochemistry. 3d ed. Prentice-Hall, Inc., Englewood Cliffs, New Jersey.

Paul, J. 1966. Cell biology: a current summary. Stanford University Press, Stanford, Calif.

Porter, K. R. and M. A. Bonneville. 1968. Fine structure of cells and tissues. 3d ed. Lea & Febiger, Philadelphia.

Robertis, E. D. P., et al. 1972. General cytology. 5th ed. W. B. Saunders Company, Philadelphia.

Rogers, H. J. and H. R. Perkins. 1968. Cell walls and membranes. E. & F. N. Spon, Ltd., London.

Swanson, C. P. 1969. The cell. 3d ed. Prentice-Hall, Inc., Englewood Cliffs, New Jersey.

The living cell. Readings from Scientific American. 1965. W. H. Freeman and Co., Publishers, San Francisco.

Zeuthen, E., ed. 1964. Synchrony in cell division and growth. Interscience Publishers, New York.

DEVELOPMENT: GROWTH AND FORM

A. Periodicals

Developmental Biology
Growth
Journal of Anatomy
Journal of Developmental Biology
Journal of Embryology and Experimental Morphology
Journal of Morphology Research on Reproduction
Symposia: Society for the Study of Development and Growth

B. General References

Allan, F. D. 1969. Essentials of human embryology. 2d ed. Oxford University Press, New York.

Arey, L. B. 1965. Developmental anatomy: a textbook and laboratory manual of embryology. 7th ed. W. B. Saunders Company, Philadelphia.

Balinsky, B. I. 1970. An introduction to embryology. 3d ed. W. B. Saunders Company, Philadelphia.

Berrell, N. J. 1971. Developmental biology. McGraw-Hill Book Co., New York.

Bonner, J. 1965. The molecular biology of development. Oxford University Press, New York.

Brachet, J. 1960. The biochemistry of development. Pergamon Press, New York.

Etkin, W. and L. Gilbert, eds. 1968. Metamorphosis: a problem of developmental biology. Appleton-Century-Crofts, New York.

Gemmell, A. R. 1969. Developmental plant anatomy. St. Martin's Press, Inc., New York.

Hanson, E. D. 1972. Animal diversity. 3d ed. Prentice-Hall, Inc., Englewood Cliffs, New Jersey.

Johnson, L. and P. E. Volpe. 1972. Patterns and experiments in developmental biology. W. C. Brown & Co., Dubuque, Iowa.

Kerr, N. S. 1971. Principles of development. 2d ed. W. C. Brown & Co., Dubuque, Iowa.

Kohn, R. R. 1971. Principles of mammalian aging. Prentice-Hall, Inc., Englewood Cliffs, New Jersey.

Maheshwari, P. 1950. An introduction to the embryology of the angiosperms. McGraw-Hill Book Co., New York.

Monroy, A. and A. A. Moscona. 1966-1970. Current topics in developmental biology. Vol. 1-5. Academic Press, New York.

Moscana, A. A., ed. 1966-1972. Current topics in developmental biology. Vol. 6. Academic Press, New York.

Patten, B. M. 1953. Early embryology of the chick. 4th ed. McGraw-Hill Book Co., New York.

_____. 1964. Foundations of embryology. 2d ed. McGraw-Hill Book Co., New York.

_____. 1968. Human embryology. 3d ed. McGraw-Hill Book Co., New York.

Romanoff, A. L. 1960. The avian embryo. The Macmillan Co., New York.

_____ and A. J. Romanoff. 1967. Biochemistry of the avian embryo. John Wiley & Sons, Inc., New York.

Rugh, R. 1953. The frog: its reproduction and development. McGraw-Hill Book Co., New York.

_____. 1971. A guide to vertebrate development. 6th ed. Burgess Pub. Co., Minneapolis, Minnesota.

_____. 1962. Experimental embryology: techniques and procedures. Burgess Pub. Co., Minneapolis, Minnesota.

_____. 1968. Mouse: its reproduction and development. Burgess Pub. Co., Minneapolis, Minnesota.

_____. 1964. Vertebrate embryology: the dynamics of development. Harcourt Brace Jovanovich, New York.

Saunders, J. W., Jr. 1970. Patterns and principles of animal development. The Macmillan Co., New York.

Sinnott, E. W. 1960. Plant morphogenesis. McGraw-Hill Book Co., New York.

Spratt, N. T., Jr. 1971. Developmental biology. Wadsworth Pub. Co., Belmont, California.

Steward, F. C. 1968. Growth and organization in plants. Addison-Wesley Pub. Co., New York.

Sussman, M. 1964. Growth and development. 2d ed. Prentice-Hall, Inc., Englewood Cliffs, New Jersey.

Waddington, C. H. 1962. Principles of embryology. Hillary House Publishers, New York.

_____. 1966. Principles of development and differentiation. The Macmillan Co., New York.

_____. 1962. New patterns in genetics and development. Columbia University Press, New York.

Weiss, P. A. 1969. Principles of development: a text in experimental embryology. Hafner Pub. Co., New York.

Wigglesworth, V. B. 1959. The control of growth and form. Cornell University Press, Ithaca, New York.

Willier, B. H. and J. M. Oppenheimer. 1964. Foundations of experimental embryology. Prentice-Hall, Inc., Englewood Cliffs, New Jersey.

ECOLOGY

A. Periodicals

American Naturalist Ecological Monographs
Ecological Society of America Bulletin
Journal of Animal Ecology
Journal of Applied Ecology
Journal of Ecology
Newsletter of Human Ecology

B. General References

Alexander, M. 1971. Microbial ecology. John Wiley & Sons, Inc., New York.

Allee, W. C., *et al.* 1949. Principles of animal ecology. W. B. Saunders Company, Philadelphia.

Andrewartha, H. G. 1961. Introduction to the study of animal populations. University of Chicago Press, Chicago.

Bardach, J. E. 1968. Harvest of the sea. Harper & Row, New York.

Bates, M. 1960. The forest and the sea. Random House, New York.

____. 1964. Man in nature. 2d ed. Prentice-Hall, Inc., Englewood Cliffs, New Jersey.

Benton, A. H. and W. E. Werner. 1972. Manual for field biology and ecology. 5th ed. McGraw-Hill Book Co., New York.

Black, C. A. 1968. Soil-plant relationships. 2d ed. John Wiley & Sons, Inc., New York.

Braun-Blanquet, J. 1966. Plant sociology: the study of plant communities. Hafner Pub. Co., New York.

Bresler, J. B. 1966. Human ecology. Addison-Wesley, Inc., Reading, Mass.

____. ed. 1968. Environments of man. Addison-Wesley, Inc., Reading, Mass.

Brock, T. D. 1966. Principles of microbial ecology. Prentice-Hall, Inc., Englewood Cliffs, New Jersey.

Burges, A. 1967. Soil biology. Academic Press, New York.

Buschbaum, R. and M. Buschbaum. 1967. Basic ecology. 8th ed. The Boxwood Press, Pittsburgh, Pennsylvania.

Carson, R. 1962. Silent spring. Houghton Mifflin Company, Boston.

Cocker, R. E. 1954. Streams, lakes, ponds. University of North Carolina Press, Chapel Hill.

Colinvaux, P. A. 1973. Introduction to ecology. John Wiley & Sons, Inc., New York.

Commoner, B. 1971. The closing circle: nature, man and technology. Alfred A. Knopf, New York.

Dasmann, R. F. 1964. Wildlife biology. John Wiley & Sons, Inc., New York.

____. 1968. Plant communities. Harper & Row, New York.

Elton, C. 1958. The ecology of invasions by animals and plants. Methuen, London.

____. 1966. The pattern of animal communities. Methuen, London.

____. 1966. Animal ecology. Barnes & Noble, Scranton, Pennsylvania.

____. 1966. The ecology of animals. Methuen, London.

Erlich, P. R. and A. H. Ehrlich. 1970. Population, resources, environment: issues in human ecology. W. H. Freeman and Co. Publishers, San Francisco.

Fried, M. and H. Broeshart. 1967. The soil-plant systems. Academic Press, New York.

Gilett, J. W., ed. 1970. The biological impact of pesticides in the environment. Oregon State University Press, Corvallis, Oregon.

Gray, T. R. G. and D. Parkinson, eds. 1968. The ecology of soil bacteria. University of Toronto Press, Toronto.

Greig-Smith, P. 1964. Quantitative plant ecology. 2d ed. Butterworth, London.

Guggisberg, C. A. W. 1970. Man and wildlife. Arco Pub. Co., New York.

Hardin, G. 1969. Population, evolution and birth control. W. H. Freeman and Co. Publishers, San Francisco.

Hesse, R., ed. 1951. Ecological animal geography. 2d ed. John Wiley & Sons, Inc., New York.

Jansen, W. A., ed. 1965. Plants and the ecosystem. Wadsworth Pub. Co., Belmont, California.

_____ and F. B. Salisbury. 1972. Botany: an ecological approach. Wadsworth Pub. Co., Belmont, California.

Kinne, O. 1970. Marine ecology: a comprehensive, integrated treatise on life in oceans and coastal waters. John Wiley & Sons, Inc., New York.

Kormondy, E. J. 1965. Readings in ecology. Prentice-Hall, Inc., Englewood Cliffs, New Jersey.

_____. 1969. Concepts in ecology. Prentice-Hall, Inc., Englewood Cliffs, New Jersey.

Leopold, A. A. 1970. A Sand County almanac. Ballantine Books, New York.

MacArthur, R. H. 1972. Geographical ecology: patterns in the distribution of species. Harper & Row, New York.

_____. 1968. Population biology and evolution. Syracuse University Press, Syracuse.

McNaughton, S. J. and L. L. Wolf. 1973. General ecology. Holt, Rinehart & Winston, New York.

Margalef, R. 1968. Perspectives in ecological theory. University of Chicago Press, Chicago.

Milne, L. and M. Milne. 1971. The nature of life: earth, plants, animals, man, and their effect on each other. Crown Publishers, New York.

_____. 1971. The arena of life, the dynamics of ecology. Doubleday, New York.

Nikolsky, G. V. 1963. The ecology of fishes. Academic Press, New York.

Odum, E. P. 1971. Fundamentals of ecology. 3d ed. W. B. Saunders Company, Philadelphia.

———. 1971. Ecosystem structure and function. Oregon State University Press, Corvallis, Oregon.

Odum, H. T. 1967. Pollution in marine ecology. John Wiley & Sons, Inc., New York.

Oosting, H. J. 1956. The study of plant communities. 2d ed. W. H. Freeman and Co., Publishers, San Francisco.

Patten, B. C., ed. 1971. Systems analysis and simulation in ecology. Academic Press, New York.

Read, C. P. 1970. Parasitism and symbiology. Ronald Press, New York.

Schmidt-Neilsen, K. 1964. Desert animals: physiological problems of heat and water. Oxford University Press, New York.

Schultz, A. M. 1971. Ecosystems and environments. Canfield Press (Harper & Row), San Francisco.

Scientific American. 1970. The biosphere. W. H. Freeman and Co. Publishers, San Francisco.

Sergerberg, O., Jr. 1971. Where have all the flowers, fishes, birds, trees, water and air gone? What ecology is all about. David McKay Co., Inc., New York.

Singer, S. F., ed. 1970. Global effects of environmental pollution. Springer-Verlag, Berlin.

Smith, R. F. and R. van den Basch, 1967. Pest control: biological, physical and selected chemical methods. Academic Press, New York.

Soundheimer, E. and J. B. Simeone, eds. 1969. Chemical ecology. Academic Press, New York.

Storer, J. H. 1956. The web of life. New American Library, New York.

———. 1968. Man in the web of life. New American Library, New York.

Venberg, F. J. and W. B. Venberg. 1970. The animal and the environment. Holt, Rinehart & Winston, New York.

Whittaker, R. H. 1970. Communities and ecosystems. The Macmillan Co., New York.

Wilson, E. O. and W. H. Bossert. 1971. A primer in population biology. Sinauer Associates, Stamford, Connecticut.

Woodwell, G. M. and H. H. Smith, eds. 1969. Diversity and stability in ecological systems. Brookhaven National Laboratory, Publication No. 22, Upton, New York.

ENTOMOLOGY

A. Periodicals

American Entomological Society Transactions
Applied Entomology and Zoology
Bulletin of Entomological Research
Canadian Entomologist
Entomological News
Entomological Review
Entomological Society of America Annals
Entomological Society of America Bulletin
Entomologist
Entomologist's Monthly Magazine
Entomologist's Record
International Journal of Insect Morphology and Embryology
International Journal of Pest Control
Journal of Economic Entomology
Journal of Insect Physiology
Journal of Research on the Lepidoptera
Mosquito News
New Entomologist
Pest Control

B. General References

Andrewes, Sir C. 1970. The lives of wasps and bees. American Elsevier Pub. Co., New York.
Askew, R. R. 1971. Parasitic insects. American Elsevier Pub. Co., New York.
Barbosa, P. and T. M. Peters, eds. 1972. Readings in entomology. W. B. Saunders Company, Philadelphia.
Beament, J. E., *et al.* 1970. Advances in insect physiology. Academic Press, New York.
Borror, D. J. and D. M. DeLong. 1970. An introduction to the study of insects. 3d ed. Holt, Rinehart & Winston, New York.

Brian, M. V. 1966. Social insect populations. Academic Press, New York.

Bursell, E. 1970. Introduction to insect physiology. Academic Press, New York.

Callahan, P. S. 1971. Insects and how they function. Holiday House, New York.

Chapman, R. F. 1969. The insects: structure and function. American Elsevier Pub. Co., New York.

———. 1971. Insects. American Elsevier Pub. Co., New York.

Clark, L. R., *et al.* 1967. Ecology of insect populations in theory and practice. Barnes & Noble, New York.

Davey, K. G. 1967. Reproduction in the insects. W. H. Freeman and Co. Publishers, San Francisco.

Dethier, V. G. 1963. Physiology of insect senses. Barnes & Noble, New York.

DuPorte, E. M. 1959. Manual of insect morphology. Van Nostrand Reinhold, New York.

Englemann, F. 1970. Physiology of insect reproduction. Pergamon Press, New York.

Gilmour, D. 1961. Biochemistry of insects. Academic Press, New York.

Jaques, H. E. 1947. How to know the insects. W. C. Brown & Co., Dubuque, Iowa.

Johnson, C. G. 1969. Migration and dispersal of insects by flight. Barnes & Noble, New York.

Linsenmaier, W. 1972. Insects of the world. McGraw-Hill Book Co., New York.

Lutz, F. E. 1948. Field book of insects of the United States and Canada. G. P. Putnam's Sons, New York.

Oldryod, H. 1960. Insects and their world. University of Chicago Press, Chicago.

———. 1970. Elements of entomology: an introduction to the study of insects. Universe Books, New York.

———. 1971. Collecting, preserving and studying insects. 2d ed. St. Martin's Press, Inc., New York.

Pfadt, R. E., ed. 1971. Fundamentals of applied entomology. 2d ed. The Macmillan Co., New York.

Ross, H. H. 1965. A textbook of entomology. 3d ed. John Wiley & Sons, Inc., New York.

Sharp, D. 1970. Insects. 2 vols. Dover Publications, New York.

Smith, D. S. 1968. Insect cells: their structure and function. S. H. Service Agency, Inc., Riverside, New Jersey.

Wheeler, W. M. 1960. Ants: their structure, development and behavior. Rev. ed. Columbia University Press, New York.

Wigglesworth, V. B. 1967. Principles of insect physiology. 6th ed. John Wiley & Sons, Inc., New York.

———. 1970. Life of insects. Universe Books, New York.

Wilson, E. O. 1971. Insect societies. Harvard University Press, Cambridge, Mass.

GENETICS AND EVOLUTION

A. Periodicals

Evolution

Genetics

Herediatas

Heredity

Journal of Genetical Research

Journal of Genetics

Journal of Heredity

Journal of Human Genetics

Journal of Molecular Biology

Journal of Evolution

Journal of Mammalogy

Annals of Human Genetics

Behavior Genetics

Nucleus

Theoretical and Applied Genetics

B. General References

Asimov, I. 1963. The genetic code. Grossman Publishers, Inc., New York.

Auerbach, C. 1965. Notes for introductory courses in genetics. Kallmann Publishing Co., Gainesville, Florida.

Barry, J. M. 1964. Molecular biology: genes and the chemical control of living cells. Prentice-Hall, Inc., Englewood Cliffs, New Jersey.

Blum, H. F. 1969. Time's arrow and evolution. Princeton University Press, Princeton, New Jersey.

Bonner, D. M. and S. E. Mills. 1964. Heredity. 2d ed. Prentice-Hall, Inc., Englewood Cliffs, New Jersey.

Burnet, M. 1971. Genes, dreams and realities. Basic Books, Inc., New York.

Colbert, E. H. 1969. Evolution of the vertebrates. 2d ed. John Wiley & Sons, Inc., New York.

Darlington, C. D. and K. R. Lewis. 1972. Chromosomes today. Hafner Pub. Co., New York.

Demerec, M. 1965. Biology of *Drosophila*. Hafner Pub. Co., New York.

——. 1947-1971. Advances in genetics. 16 vols. of 25. Academic Press, New York.

Dobzhansky, T. 1971. Genetics of the evolutionary process. Columbia University Press, New York.

——. 1962. Mankind evolving: the evolution of the human species. Yale University Press, New Haven, Connecticut.

——., *et al.* 1969-1972. Evolutionary biology. 6 vols. Appleton-Century-Crofts, New York.

Esser, K. and R. Kuenen. 1967. Genetics of fungi. Springer-Verlag, New York.

Fincham, J. R. S. and P. Day. 1971. Fungal genetics. 3d ed. F. A. Davis Co., Philadelphia.

Gardner, E. J. 1972. Principles of genetics. 4th ed. John Wiley & Sons, Inc., New York.

Halstead, L. B. 1969. The pattern of vertebrate evolution. W. H. Freeman and Co. Publishers, San Francisco.

Harris, H. 1970. Frontiers of biology: the principles of human biochemical genetics. American Elsevier Pub. Co., New York.

Hayes, W. 1969. Genetics of bacteria and their viruses. 2d ed. John Wiley & Sons, Inc., New York.

Kalmus, H. 1964. Genetics. Natural History Press, New York.

King, R. C. 1965. Genetics. 2d ed. Oxford University Press, New York.

——. 1972. Dictionary of genetics. 2d ed. Oxford University Press, New York.

Lerner, I. M. 1968. Heredity, evolution and society. W. H. Freeman and Co. Publishers, San Francisco.

Levine, L. 1971. Papers on genetics: a book of readings. C. V. Mosby Company, St. Louis.

Lewin, B. M. 1970. The molecular basis of gene expression. John Wiley & Sons, Inc., New York.

Lewis, K. R. and J. Bernard. 1970. The organization of heredity. American Elsevier Pub. Co., New York.

McKusick, V. A. 1969. Human genetics. 2d ed. Prentice-Hall, Inc., Englewood Cliffs, New Jersey.

Mayr, E. 1970. Populations, species, and evolution. Harvard University Press, Cambridge, Massachusetts.

Merrell, D. J. 1962. Evolution and genetics. Holt, Rinehart & Winston, New York.

Moore, J. A. 1972. Heredity and development. 2d ed. Oxford University Press, New York.

_____. 1972. Readings in heredity and development. Oxford University Press, New York.

Pedder, I. J. and E. G. Wynne. 1972. Genetics: a basic guide. Hutchinson Educational, London.

Peters, J. A. 1961. Classic papers in genetics. Prentice-Hall, Inc., Englewood Cliffs, New Jersey.

Provine, W. B. 1971. The origins of theoretical population genetics. University of Chicago Press, Chicago.

Raper, J. R. 1966. Genetics of sexuality in higher fungi. Ronald Press, New York.

Ross, H. H. 1962. Synthesis of evolutionary theory. Prentice-Hall, Inc., Englewood Cliffs, New Jersey.

Scientific American. 1970. Facets of genetics. W. H. Freeman and Co. Publishers, San Francisco.

Silverstein, A. and V. Silverstein. 1972. The code of life. Antheneum, New York.

Simpson, G. G. 1967. Geography of evolution. G. P. Putnam's Sons, New York.

_____. 1967. Meaning of evolution: a study of the history of life and of its significance for man. Yale University Press, New Haven, Connecticut.

Srb, A. M., R. D. Owen and R. S. Edger. 1965. General genetics. 2d ed. W. H. Freeman and Co. Publishers, San Francisco.

Stahl, F. W. 1969. The mechanics of inheritance. 2d ed. Prentice-Hall, Inc., Englewood Cliffs, New Jersey.

Stebbins, G. L. 1971. Processes of organic evolution. 2d ed. Prentice-Hall, Inc., Englewood Cliffs, New Jersey.

Strickberger, M. W. 1968. Genetics. The Macmillan Co., New York.

Wallace, B. 1966. Chromosomes: giant molecules and evolution. W. W. Norton, Inc., New York.

——. 1968. Topics in population genetics. W. W. Norton, Inc., New York.

——. 1972. Genetics, evolution, race, radiation biology: essays in social biology. Prentice-Hall, Inc., Englewood Cliffs, New Jersey.

Watson, J. D. 1965. Molecular biology of the gene. W. A. Benjamin, Inc., New York.

——. 1968. The double helix. Antheneum, New York.

Winchester, A. M. 1972. Genetics: a survey of the principles of heredity. 4th ed. Houghton Mifflin Company, Boston.

——. 1971. Human genetics. Charles E. Merrill Publishing Co., Columbus, Ohio.

Woese, C. R. 1967. The genetic code: the molecular basis for genetic expression. Harper & Row, New York.

KEYS: MANUALS FOR IDENTIFICATION

A. General

Breed, R. S., E. Murray and N. Smith, eds. 1967. Bergey's manual of determinative bacteriology. 7th ed. Williams & Wilkins, Co., Baltimore.

Edmondson, W. T., *et al.* 1959. Freshwater biology. 2d ed. John Wiley & Sons, Inc., New York.

Gibbs, B. M. and F. A. Skinner, eds. 1968. Identification methods for microbiologists. Academic Press, New York.

Jaques, H. E. 1946. Living things—how to know them. W. C. Brown Co., Dubuque, Iowa.

Morgan, A. H. 1930. Field book of ponds and streams. G. P. Putnam's Sons, New York.

Needham, J. G. and P. R. Needham. 1969. A guide to the study of freshwater biology. 5th rev. ed. Holden-Day, Inc., San Francisco.

Roso-MacDonald, M., ed. 1971. The world wildlife guide. Viking Press, Inc., New York.

B. Animals

Booth, E. S. 1971. How to know the mammals. 2d ed. W. C. Brown Co., Dubuque, Iowa.

Borror, D. J. and R. E. White. 1970. A field guide to the insects of America north of Mexico. Houghton Mifflin Co., Boston.

Burt, W. H. and R. P. Grossenheider. 1964. A field guide to the mammals. 2d ed. Houghton Mifflin Co., Boston.

Comstock, J. H. 1948. The spider book. Comstock Pub. Co., Ithaca, New York.

Conant, R. 1958. A field guide to reptiles and amphibians. Houghton Mifflin Co., Boston.

Chu, H. F. 1949. How to know the immature insects. W. C. Brown Co., Dubuque, Iowa.

Eddy, S. 1970. How to know the freshwater fishes. 2d ed. W. C. Brown Co., Dubuque, Iowa.

Erlich, P. R. and A. H. Erlich. 1961. How to know the butterflies. W. C. Brown Co., Dubuque, Iowa.

Jahn, T. L. 1949. How to know the protozoa. W. C. Brown Co., Dubuque, Iowa.

Helfer, J. R. 1963. How to know the grasshoppers. W. C. Brown Co., Dubuque, Iowa.

Jaques, H. E. 1947. How to know the insects. W. C. Brown Co., Dubuque, Iowa.

_____. 1951. How to know the beetles. W. C. Brown Co., Dubuque, Iowa.

_____. 1947. How to know the land birds. W. C. Brown Co., Dubuque, Iowa.

_____ and R. Olliver. 1960. How to know the water birds. W. C. Brown Co., Dubuque, Iowa.

Kaston, B. I. and E. Kaston. 1972. How to know the spiders. W. C. Brown Co., Dubuque, Iowa.

Klots, A. B. 1951. A field guide to the butterflies of North America, east of the Great Plains. Houghton Mifflin Co., Boston.

_____ and E. B. Klots. 1971. Insects of North America. Doubleday & Company, Inc., New York.

Kudo, R. 1971. Protozoology. 5th ed. Charles C. Thomas, Springfield, Illinois.

Lutz, F. E. 1948. Field book of insects of the United States and Canada. G. P. Putnam's Sons, New York.

Morris, P. A. 1966. A field guide to the shells of our Atlantic and Gulf Coast. Houghton Mifflin Co., Boston.

——. 1966. A field guide to the shells of the Pacific Coast and Hawaii. Houghton Mifflin Co., Boston.

Pennak, R. W. 1953. Freshwater invertebrates of the United States. Ronald Press Co., New York.

Peterson, R. T. 1968. A field guide to the birds. Houghton Mifflin Co., Boston.

Pimentel, R. A. 1967. Invertebrate identification manual. Van Nostrand Reinhold, New York.

Smith, H. M. 1946. Handbook of lizards. Comstock Pub. Co., Ithaca, New York.

C. Plants

Baerg, H. J. 1955. How to know the western trees. W. C. Brown Co., Dubuque, Iowa.

Bell, R. C. 1967. Plant variation and classification. Wadsworth Pub. Co., Belmont, California.

Bold, H. C. 1964. The plant kingdom. Prentice-Hall, Inc., Englewood Cliffs, New Jersey.

Conard, H. S. 1956. How to know the mosses and liverworts. W. C. Brown Co., Dubuque, Iowa.

Cuthbert, M. J. 1948. How to know the fall flowers. W. C. Brown Co., Dubuque, Iowa.

——. 1948. How to know the spring flowers. W. C. Brown Co., Dubuque, Iowa.

Dawson, G. Y. 1963. How to know the seaweeds. W. C. Brown Co., Dubuque, Iowa.

——. 1963. How to know the cacti. W. C. Brown Co., Dubuque, Iowa.

Harlow, W. M. 1959. Trees. Dover Publications, New York.

Jaques, H. E. 1958. How to know the economic plants. W. C. Brown Co., Dubuque, Iowa.

——. 1946. How to know the trees. W. C. Brown Co., Dubuque, Iowa.

——. 1959. How to know the weeds. W. C. Brown Co., Dubuque, Iowa.

——. 1948. Plant families—how to know them. W. C. Brown Co., Dubuque, Iowa.

Montgomery, F. H. 1964. Weeds of the northern United States and Canada. Ryerson Press, Toronto.

_____. 1965. Native wild plants of northern United States and eastern Canada. Frederick Warne & Co., Inc., New York.

_____. 1971. Trees of the northern United States and Canada. Frederick Warne & Co., Inc., New York.

Nelson, R. A. 1970. Plants of Rocky Mountain National Park. Rocky Mountain Nature Association.

Petrides, G. A. 1972. A field guide to trees and shrubs. 2d ed. Houghton Mifflin Co., Boston.

Pohl, R. W. 1968. How to know the grasses. 2d ed. W. C. Brown Co., Dubuque, Iowa.

Prescott, G. W. 1970. How to know the freshwater algae. 2d ed. W. C. Brown Co., Dubuque, Iowa.

Smith, G. M. 1950. Freshwater algae of the United States. 2d ed. McGraw-Hill Book Co., New York.

Torrey, J. and A. Gray. 1968. Flora of North America. Hafner Pub. Co., New York.

Tribe, I. 1970. The plant kingdom. Grossett & Dunlap, New York.

MICROBIOLOGY

A. Periodicals

Advances in Virus Research

Annual Reviews of Microbiology

Applied Microbiology

Bacteriological Reviews

Currents in Modern Biology

International Journal of Systematic Bacteriology

Journal of Applied Bacteriology

Journal of General and Applied Microbiology

Journal of General Microbiology

Journal of General Virology

Journal of Medical Microbiology

Journal of Protozoology

Journal of Virology
Microbiology
Virology
Virus

B. General References

Adelberg, E. A., ed. 1966. Papers on bacterial genetics. 2d ed. Little, Brown & Co., Boston.

Ainsworth, G. C. and A. S. Sussman, eds. 1965-1968. The fungi: an advanced treatise. 3 vols. Academic Press, New York.

Alexander, M. 1971. Microbial ecology. John Wiley & Sons, Inc., New York.

Alexopoulos, C. J. and H. C. Bold. 1967. Algae and fungi. The Macmillan Co., New York.

Bailey, W. R. and E. G. Scott. 1970. Diagnostic microbiology. 3d ed. C. V. Mosby Co., St. Louis.

Bisset, K. A. 1964. Bacteria. 3d ed. Williams & Wilkins Co., Baltimore.

Braun, W. 1965. Bacterial genetics. 2d ed. W. B. Saunders Company, Philadelphia.

Brock, T. D. 1961. Milestones in microbiology. Prentice-Hall, Inc., Englewood Cliffs, New Jersey.

——. 1966. Principles of microbial ecology. Prentice-Hall, Inc., Englewood Cliffs, New Jersey.

——. 1969. Biology of Microorganisms. Prentice-Hall, Inc., Englewood Cliffs, New Jersey.

Burrows, W., R. J. Porter and J. W. Moulder. 1968. Textbook of microbiology. 19th ed. 2 vols. W. B. Saunders Company, Philadelphia.

Carpenter, P. L. 1972. Microbiology. 3d ed. W. B. Saunders Company, Philadelphia.

Davis, B. D., et al. 1967. Microbiology. Harper & Row, New York.

Fincham, J. R. S. and P. R. Day. 1963. Fungal genetics. F. A. Davis Co., Philadelphia.

Fogg, G. E. 1965. Algal cultures and phytoplankton ecology. University of Wisconsin Press, Madison.

Fraser, D. 1967. Viruses and molecular biology. The Macmillan Co., New York.

Frobisher, M. 1968. Fundamentals of microbiology. 8th ed. W. B. Saunders Company, Philadelphia.

Gray, T. R. G. and D. Parkinson, eds. 1968. The ecology of soil bacteria. Liverpool University Press, Liverpool.

Hahon, N., ed. 1964. Selected papers on virology. Prentice-Hall, Inc., Englewood Cliffs, New Jersey.

Hayes, W. 1968. The genetics of bacteria and their viruses. 2d ed. Blackwell Scientific Publishers, Oxford, England.

Kudo, R. 1971. Protozoology. 5th ed. Charles C. Thomas, Springfield, Illinois.

Luria, S. E. and J. E. Darnell, Jr. 1967. General virology. John Wiley & Sons, Inc., New York.

Mackinnon, D. L. and R. S. J. Hawes. 1961. An introduction to the study of protozoa. Oxford University Press, New York.

Maramorosch, K. and H. Koprowski, eds. 1967-1968. Methods in virology. 4 vols. Academic Press, New York.

Pelczar, M. J. and R. D. Reid. 1972. Microbiology. 3d ed. McGraw-Hill Book Co., New York.

Poindexter, J. S. 1971. Microbiology: an introduction to the protists. The Macmillan Co., New York.

Pramer, D. and E. L. Schmidt. 1965. Experimental soil microbiology. Burgess Pub. Co., Minneapolis.

Rose, A. H. 1968. Chemical microbiology. 2d ed. Butterworth, London.

Salle, A. J. 1967. Fundamental principles of bacteriology. 6th ed. McGraw-Hill Book Co., New York.

Salton, M. R. J. 1964. The bacterial cell wall. American Elsevier Pub. Co., New York.

Smith, K. M. 1967. Insect virology. Academic Press, New York.

_____. 1968. Plant viruses. 4th ed. John Wiley & Sons, Inc., New York.

Smith, P. F. 1971. The biology of mycoplasms. Academic Press, New York.

Stanier, R. Y., M. Doudoroff and E. A. Adelberg. 1970. The microbial world. Prentice-Hall, Inc., Englewood Cliffs, New Jersey.

Stent, G. S. 1963. Molecular biology of bacterial viruses. W. H. Freeman and Co. Publishers, San Francisco.

Webster, J. 1970. Introduction to fungi. Cambridge University Press, New York.

Weidel, W. 1959. Virus. University of Michigan Press, Ann Arbor.

APPENDIX B

PHYSIOLOGY

A. Periodicals

American Journal of Physiology
Comparative Biochemistry and Physiology
Human Development
Journal of American Morphology and Physiology
Journal of Applied Physiology
Journal of Cellular Physiology
Journal of General Physiology
Journal of Neurophysiology
Journal of Physiology
Physiological Review
Physiologist
Physiology and Behavior
Physiological Zoology

B. General References

American Physiological Society. 1962-1967. Handbook of physiology. 3 vols. Williams & Wilkins Co., New York.

Bell, G. H. 1969. Textbook of physiology and biochemistry. 7th ed. Williams & Wilkins Co., New York.

Dowben, R. M. 1969. General physiology: a molecular approach. Harper & Row, New York.

Giese, A. 1968. Cell physiology. 3d ed. W. B. Saunders Company, Philadelphia.

Lehninger, A. L. 1971. Bioenergetics: the molecular basis of biological energy transformation. 2d ed. W. A. Benjamin, Inc., New York.

Rothstein, H. 1971. General physiology: the cellular and molecular basis. Xerox College Pub. Co., Waltham, Mass.

Ruch, T. C. and H. D. Patton, eds. 1965. Physiology and biophysics. 19th ed. W. B. Saunders Company, Philadelphia.

C. Animals

Anthony, C. P. 1972. Structure and function of the body. Rev. 4th ed. C. V. Mosby Co., St. Louis.

Best, C. H. and N. B. Taylor. 1973. Living body: a text in human physiology. 5th ed. Holt, Rinehart & Winston, New York.

Bishop, M., ed. 1971. Advances in reproductive physiology. Academic Press, New York.

Carlson, A. J. and V. Johnson. 1961. The machinery of the body. 5th ed. University of Chicago Press, Chicago.

Cloudsley-Thompson, J. L. 1961. Rhythmic activity in animal physiology and behavior. Academic Press, New York.

Gordon, M. S. 1972. Animal physiology: principles and adaptations. The Macmillan Co., New York.

Gray, J. 1953. How animals move. Cambridge University Press, London.

Mellon, De F., Jr. 1968. The physiology of sense organs. W. H. Freeman and Co. Publishers, San Francisco.

Prosser, C. L. and F. A. Brown, Jr. 1961. Comparative animal physiology. 2d ed. W. B. Saunders Company, Philadelphia.

Roberts, T. D. M. 1966. Basic ideas in neurophysiology. Appleton-Century-Crofts, New York.

Schmidt-Neilsen, K. 1964. Desert animals: physiological problems of heat and water. Claredon Press, Oxford, England.

____. 1970. Animal physiology. 3d ed. Prentice-Hall, Inc., Englewood Cliffs, New Jersey.

Turner, C. D. and J. T. Bagnara. 1971. General endocrinology. 5th ed. W. B. Saunders Company, Philadelphia.

D. Plants

Audus, L. J. 1960. Plant growth substances. 2d ed. John Wiley & Sons, Inc., New York.

Bonner, J. F. and A. W. Galston. 1952. Principles of plant physiology. W. H. Freeman and Co. Publishers, San Francisco.

Devlin, R. M. 1966. Plant physiology. Reinhold Pub. Co., New York.

____ and A. V. Barker. 1971. Photosynthesis. Van Nostrand Reinhold, New York.

Epstein, E. 1972. Mineral nutrition of plants: principles and perspectives. John Wiley & Sons, Inc., New York.

Hillman, W. S. 1962. The physiology of flowering. Holt, Rinehart & Winston, New York.

———. 1970. Papers on plant physiology. Holt, Rinehart & Winston, New York.

Kamen, M. D. 1963. Primary processes in photosynthesis. Academic Press, New York.

Kozlowski, T. T. 1964. Water metabolism in plants. Harper & Row, New York.

Loomis, W. 1960. Encyclopedia of plant physiology. Springer-Verlag, Berlin.

McElroy, W. D. and B. Glass, eds. 1954. Light and life. Johns Hopkins Press, Baltimore.

O'hEocha, C. 1962. Physiology and biochemistry of algae. Academic Press, New York.

Rabinowitch, E. 1969. Photosynthesis. John Wiley & Sons, Inc., New York.

Salisbury, F. 1971. The biology of flowering. Natural History, Garden City, New York.

——— and C. Ross. 1969. Plant physiology. Wadsworth Pub. Co., Belmont, California.

Steward, W. D. 1966. Nitrogen fixation in plants. Anthenuem, New York.

Stills, W. 1969. An introduction to the principles of plant physiology. 3d ed. Barnes & Noble, New York.

PLANTS

A. Periodicals

American Journal of Botany
Annual Reviews of Plant Physiology
Botanical Gazette
Botanical Review
Bryologist
Bulletin of the Torrey Botanical Club
Cactus and Succulent Journal

Canadian Journal of Botany
Canadian Journal of Plant Science
Economic Botany
Frontiers of Plant Science
Journal of Experimental Botany
Lichenologist
Mycologia
Phytomorphology
Plant and Gardens
Plant and Soil
Plant Disease Reporter
Plant Physiology
Plant Pathology
Plant Science Bulletin

B. General References

Andrews, H. N., Jr. 1961. Studies in paleobotany. John Wiley & Sons, Inc., New York.

Arnett, R. H. and D. C. Braungart. 1970. An introduction to plant biology. C. V. Mosby Co., St. Louis.

Audus, L. J. 1964. Physiology and biochemistry of herbicides. Academic Press, New York.

Bell, P. R. and C. L. Woodcock. 1972. The diversity of green plants. Addison-Wesley Pub. Co., New York.

Bierhorst, D. W. 1971. Morphology of vascular plants. The Macmillan Co., New York.

Bold, H. C. 1967. Morphology of plants. Harper & Row, New York.

_____. 1970. Plant kingdom. 2d ed. Prentice-Hall, Inc., Englewood Cliffs, New Jersey.

Bonner, J. F. 1966. Plant biochemistry. Academic Press, New York.

_____ and A. W. Galston. 1952. Principles of plant physiology. W. H. Freeman and Co. Publishers, San Francisco.

Britton, N. L. and A. Brown. 1967. Illustrated flora of the northern United States and Canada. 2d ed. 3 vols. Peter Smith, Gloucester, Massachusetts.

Clarkson, Q. D. 1961. Handbook of field botany. Binfords, New York.

Cronquist, A. 1971. Introductory botany. 2d ed. Harper & Row, New York.

Dodd, J. D. 1962. Form and function in plants. Iowa State University Press, Ames, Iowa.

Eames, A. J. 1961. Morphology of the angiosperms. McGraw-Hill Book Co., New York.

Esau, K. 1960. Anatomy of seed plants. John Wiley & Sons, Inc., New York.

————. 1961. Plants, viruses, and insects. Harvard University Press, Cambridge, Massachusetts.

————. 1965. Plant anatomy. 2d ed. John Wiley & Sons, Inc., New York.

Foster, A. S. and E. M. Gifford. 1959. Comparative morphology of vascular plants. 2d ed. John Wiley & Sons, Inc., New York.

Fuller, H. J. and O. Tippo. 1954. College botany. Holt, Rinehart & Winston, New York.

Grant, V. 1971. Plant speciation. Columbia University Press, New York.

Gray, A. 1950. Manual of botany. 8th ed. Van Nostrand Reinhold, New York.

————. 1970. Elements of botany. Arno Press, New York.

Greulach, V. A. and J. E. Adams. 1962. Plants: introductory investigations in botany. John Wiley & Sons, Inc., New York.

Hale, M. E., Jr. 1970. The biology of the lichens. American Elsevier Pub. Co., New York.

Halvorson, H. O., ed. 1961. Spores. Burgess Pub. Co., Minneapolis, Minn.

Haupt, A. W. 1953. Plant morphology. McGraw-Hill Book Co., New York.

Lawrence, W. J. C. 1968. Plant breeding. St. Martin's Press, Inc., New York.

Northern, H. and R. Northern. 1970. Ingenious kingdom: the remarkable world of plants. Prentice-Hall, Inc., Englewood Cliffs, New Jersey.

Raven, P. and H. Curtis. 1970. Biology of plants. Worth Publishers, Inc., New York.

Scagel, R. F. 1965. Evolutionary survey of the plant kingdom. Wadsworth Pub. Co., New York.

————. 1969. Plant diversity: an evolutionary approach. Wadsworth Pub. Co., New York.

Scientific American Eds. 1957. Plant life. Simon & Schuster, New York.

Sinnott, E. W. and K. S. Wilson. 1963. Botany: principles and problems. 6th ed. McGraw-Hill Book Co., New York.

Smith, G. M. 1955. Cryptogamic botany. 2d ed. 2 vols. McGraw-Hill Book Co., New York.

Sporne, K. R. 1967. Morphology of the gymnosperms: the structure and evolution of primitive seed plants. Hutchinson Scientific Press, Baltimore.

Steward, F. C. and A. D. Krikorain. 1971. Plants, chemicals and growth. Academic Press, New York.

United States Department of Agriculture. 1952. Yearbook of agriculture. Plant diseases. U. S. Government Printing Office, Washington, D. C.

Van Der Pijl, L. 1972. Principles of dispersal in higher plants. Springer-Verlag, New York.

Weier, T. C., C. R. Stocking and M. G. Barbour. 1970. Botany: an introduction to plant biology. 4th ed. John Wiley & Sons, Inc., New York.

Went, F. W. 1957. Experimental control of plant growth. Ronald Press, New York.

RESEARCH: METHODS AND TECHNIQUES

Allard, R. W. 1960. Principles of plant breeding. John Wiley & Sons, Inc., New York.

Allen, R. M. 1958. Photomicrography. 2d ed. D. Van Nostrand Co., Princeton, New Jersey.

Aronoff, S. 1967. Techniques of radiobiochemistry. Hafner Pub. Co., New York.

Bailey, N. T. J. 1959. Statistical methods in biology. John Wiley & Sons, Inc., New York.

———. 1967. Mathematical approach to biology and medicine. John Wiley & Sons, Inc., New York.

Baker, F. J. 1967. Handbook of bacteriological technique. 2d ed. Appleton-Century-Crofts, New York.

Barnes, H. 1968. Oceanography and marine biology: a book of techniques. Hafner Pub. Co., New York.

Beveridge, W. I. B. 1957. The art of scientific investigation. Rev. ed. W. W. Norton, New York.

Biological Sciences Curriculum Study. 1970. Biology teachers handbook. 2d ed. John Wiley & Sons, Inc., New York.

Block, R. J., E. L. Durrum and G. Zweig. 1958. A manual of paper chromatography and paper electrophoresis. 2d ed. Academic Press, New York.

Conn, H. J., *et al.* 1960. Staining procedures. 3d ed. Williams & Wilkins Co., Baltimore.

Davenport, H. A. 1960. Histological and histochemical techniques. W. B. Saunders Company, Philadelphia.

Demerec, M. and B. P. Kaufmann. 1961. *Drosophila* guide. Carnegie Institute of Washington, Washington, D. C.

Dixon, W. J. and F. J. Massey. 1969. Introduction to statistical analysis. 3d ed. McGraw-Hill Book Co., New York.

Downie, N. M. and R. W. Heath. 1970. Basic statistical methods. 3d ed. Harper & Row, New York.

Duddington, C. L. 1960. Practical microscopy. Pitman Publishing Corporation, New York.

Emmel, V. M. and E. V. Cowdry. 1964. Laboratory technique in biology and medicine. Williams & Wilkins Co., Baltimore.

Ferris, E. J., ed. 1950. The care and breeding of laboratory animals. John Wiley & Sons, Inc., New York.

Fisher, R. A. and F. Yates. 1964. Statistical tables for biological, agricultural and medical research. 6th ed. Hafner Pub. Co., New York.

Freud, J. E. 1967. Modern elementary statistics. 3d ed. Prentice-Hall, Inc., Englewood Cliffs, New Jersey.

Galigher, A. E. and E. N. Kozloff. 1971. Essentials of practical microtechnique. 2d ed. Lea & Febiger, Philadelphia.

Gander, R. 1969. Photomicrographic technique for medical and biological scientists. Hafner Pub. Co., New York.

Goldstein, P. 1957. How to do an experiment. Harcourt, Brace & World, New York.

Gordon, A. H. and J. E. Eastoe. 1964. Practical chromatographic techniques. Van Nostrand Reinhold, New York.

Gray, P. 1954. The microtomist's formulary and guide. Blakiston Co., New York.

____. 1964. Handbook of basic microtechnique. 3d ed. McGraw-Hill Book Co., New York.

Guyer, M. R. 1953. Animal micrology: practical exercises in zoological microtechnique. 5th ed. University of Chicago Press, Chicago.

Hall, C. E. 1966. Introduction to electron microscopy. 2d ed. McGraw-Hill Book Co., New York.

Hartmann, H. T. and D. E. Kester. 1968. Plant propagation: principles and practices. 2d ed. Prentice-Hall, Inc., Englewood Cliffs, New Jersey.

Heftmann, E., ed. 1967. Chromatography. 2d ed. Van Nostrand Reinhold, New York.

Humason, G. L. 1972. Animal tissue techniques. 3d ed. W. H. Freeman and Co. Publishers, San Francisco.

Jones, R. M., ed. 1940. McClung's handbook of microscopic techinque for workers in animal and plant tissues. 3d ed. Harper & Row, New York.
____. 1966. Basic microscopic techniques. Based on Guyer's animal micrology. University of Chicago Press, Chicago.

Levine, M. and H. W. Schoeline. 1960. A compilation of culture media for the cultivation of microorganisms. Williams & Wilkins Co., Baltimore.
____. 1954. Laboratory technique in bacteriology. 3d ed. The Macmillan Co., New York.

Lotka, A. J. 1957. Elements of mathematical biology. Dover Publications, New York.

Loveland, R. P. 1970. Photomicrography: a comprehensive treatise. 2d ed. 2 vols. John Wiley & Sons, Inc., New York.

Machlis, L. and J. G. Torrey. 1956. Plants in action—a laboratory manual of plant physiology. W. H. Freeman and Co. Publishers, San Francisco.

Meek, G. A. 1970. Practical electron microscopy for biologists. Wiley-Interscience, New York.

Miller, D. F. and G. W. Blaydes. 1962. Methods and materials for teaching the biological sciences. 2d ed. McGraw-Hill Book Co., New York.

Moore, C. B. 1954. Book of wild pets. Charles T. Branford, Newton Center, Mass.

National Research Council. 1954. Handbook of laboratory animals. National Academy of Sciences, Washington, D. C.

Needham, J. G. 1937. Culture methods of invertebrate animals. Dover Publications, New York.

Pantin, C. F. 1960. Notes on microscopical technique for zoologists. Cambridge University Press, New York.

Parker, R. C. 1962. Methods of tissue culture. 3d ed. Harper & Row, New York.

Paul, J. R. 1970. Cell and tissue culture. 4th ed. Williams & Wilkins Co., Baltimore.

Preece, A. 1972. Manual for histologic technicians. 3d ed. Little, Brown & Co., Boston.

Rugh, R. 1962. Experimental embryology: techniques and procedures. Burgess Pub. Co., Minneapolis.

Siegel, B. M., ed. 1964. Modern developments in electron microscopy. Academic Press, New York.

Slayter, E. M. 1970. Optical methods in biology. Wiley-Interscience, New York.

Snedecor, G. W. and E. W. G. Cochran. 1967. Statistical methods. 6th ed. Iowa State University Press, Ames, Iowa.

Society of American Bacteriologists. 1957. Manual of microbial methods. McGraw-Hill Book Co., New York.

Stehli, G. J. 1970. The microscope and how to use it. Dover Publications, New York.

Weesner, F. M. 1960. General zoological microtechniques. Williams & Wilkins Co., Baltimore.

Welch, P. S. 1948. Limnological methods. McGraw-Hill Book Co., New York.

White, P. R. 1963. The cultivation of animal and plant cells. 2d ed. Ronald Press, New York.

Zweifel, F. W. 1961. A handbook of biological illustration. University of Chicago Press, Chicago.

APPENDIX C — BSCS PUBLICATIONS

BSCS INTRODUCTORY BIOLOGY COURSES—approaches to science as inquiry.

Biological Science: An Inquiry into Life (BSCS Yellow Version)
Teacher's Guide and Resource Book of Test Items
Harcourt Brace Jovanovich, Inc., New York, N. Y. 10017

Biological Science: Molecules to Man (BSCS Blue Version)
Teacher's Guide
Houghton Mifflin Company, Boston, Mass. 02107

Resource Book of Test Items
Biological Sciences Curriculum Study, Department BEM
P. O. Box 930, Boulder, Colo. 80302

Biological Science: An Ecological Approach (BSCS Green Version)
Teacher's Edition
Rand McNally & Company, Chicago, Ill. 60680

Resource Book of Test Items
Biological Sciences Curriculum Study, Department BEM
P. O. Box 930, Boulder, Colo. 80302

Biological Science: Invitations to Discovery
Teacher's Guide
Holt, Rinehart and Winston, Inc., New York, N. Y. 10017

Biological Science: Patterns and Processes
Teacher's Edition
Holt, Rinehart and Winston, Inc., New York, N. Y. 10017

BSCS SECOND COURSE—biological investigation for students who have completed an introductory course.

Biological Science: Interaction of Experiments and Ideas
Teacher's Guide
Prentice-Hall, Inc., Englewood Cliffs, N. J. 07632

Achievement Tests
Biological Sciences Curriculum Study
P. O. Box 930, Boulder, Colo. 80302

BSCS COLLEGE BIOLOGY

Minicourses in Biology—Student Guides, Instructor's Guides, Tapescripts, Film Loops in an audiotutorial program
W. B. Saunders Company, Philadelphia, Pa. 19105

Biology: Self, Community, and Society (tentative title—book in preparation)
Student Manual, Instructor's Manual, MAGAbacks
Little, Brown and Company, Boston, Mass. 02109

BSCS RESEARCH PROBLEMS—research suggested for students by practicing research scientists.

Research Problems in Biology—Series 1, 2, and 3
Oxford University Press, New York, N. Y. 10016

BSCS MODULAR BIOLOGY

Investigating Your Environment—Student Handbook, Environmental Resource Papers, Teacher's Handbook
Addison—Wesley Publishing Company, Menlo Park, Calif. 94025

Energy and Human Affairs (in preparation)
Publisher to be announced

BSCS ELEMENTARY, MIDDLE SCHOOL, AND JUNIOR HIGH PROGRAMS

BSCS—Lippincott Elementary School Sciences Program
J. B. Lippincott Company, Philadelphia, Pa. 19105

Human Sciences Program
Publisher to be announced

BSCS PROGRAMS FOR THE EDUCABLE MENTALLY HANDICAPPED

Me Now
Me and My Environment
Hubbard Scientific Company, Northbrook, Ill. 60062

BSCS AUDIO–VISUAL PROGRAMS

Films—16mm color, sound
    Observing Behavior
    Parents
    A Member of Society
    What's Mine Is Mine
    Language without Words
    The Tragedy of the Commons
    Interview with Garrett Hardin
    Energy to Burn
    Elephant Seals
        BFA Educational Media, Santa Monica, Calif. 90404

Films—16mm color, sound, or sound-slide series
    Projections for the Future—A Growth Model
    Projections for the Future—A Behavior Model
    Projections for the Future—A Humanist Model
        Crystal Productions, Aspen, Colo. 81611

Films—Super 8mm film loops
    Single Topic Inquiry Films (40 titles)
        Harcourt Brace Jovanovich, Inc., New York, N. Y. 10017
        Rand McNally & Company, Chicago, Ill. 60680
        Hubbard Scientific Company, Northbrook, Ill. 60062

Sound-Slide and Slide Series
    An Inquiry into the Origin of Man: Science and Religion
    The New Genetics: Rights and Responsibilities
        Science and Mankind, Inc., P. O. Box 930, Boulder, Colo. 80302

    BSCS Inquiry Slides
        Harcourt Brace Jovanovich, Inc., New York, N. Y. 10017

    Investigations in Life Science: Man and Nature
        Crystal Productions, Aspen, Colo. 81611

Stereo Pictures
    Biology in Three Dimensions
        Hubbard Scientific Company, Northbrook, Ill. 60062

BSCS LABORATORY BLOCKS, PAMPHLETS, BOOKS ON SOCIAL ISSUES AND BIOLOGICAL TOPICS, BULLETINS, SPECIAL PUBLICATIONS, AND NEWSLETTER

Laboratory Blocks (12 titles)
D. C. Heath & Company, Lexington, Mass. 02173

Laboratory Block (1 title—Radiation and Its Use in Biology)
Biological Sciences Curriculum Study
P. O. Box 930, Boulder, Colo. 80302

Pamphlets and Patterns of Life Series (34 titles)
Biological Sciences Curriculum Study, Department BEM
P. O. Box 930, Boulder, Colo. 80302

Population Genetics—A Self-Instructional Program
General Learning Corporation, Silver Burdett Division,
Morristown, N. J. 07960

Science and Society Series (10 titles)
Topics in Biological Science (3 titles)
    The Bobbs-Merrill Company, Indianapolis, Ind. 46268

Bulletins and Special Publications (4 titles in print)
    Biological Sciences Curriculum Study
    P. O. Box 930, Boulder, Colo. 80302

Planning Curriculum Development (with Examples from Projects for
the Mentally Retarded)
    Biological Sciences Curriculum Study
    P. O. Box 930, Boulder, Colo. 80302

Biology Teachers' Handbook
    John Wiley & Sons, Inc., New York, N. Y. 10016

BSCS Newsletter (issued four times a year)
    Biological Sciences Curriculum Study
    P. O. Box 930, Boulder, Colo. 80302